www.wieser-verlag.com

Progress in Mathematics
Volume 147

Series Editors
Hyman Bass
Joseph Oesterlé
Alan Weinstein

Fausto Di Biase

Fatou Type Theorems
Maximal Functions
and Approach Regions

Birkhäuser
Boston • Basel • Berlin

Fausto Di Biase
Department of Mathematics
Princeton University
Princeton, NJ 08540

Currently at:
University Roma-Tor Vergata
Dip. Matematica
00133 Rome, Italy

Library of Congress Cataloging-in-Publication Data

Di Biase, Fausto, 1962-
  Fatou type theorems : maximal functions and approach regions /
Fausto Di Biase.
      p.   cm. -- (Progress in mathematics ; v. 147)
      Includes bibliographical references (p.   -   ) and index.
      ISBN 0-8176-3976-4 (acid-free paper). -- ISBN 3-7643-3976-4 (Basel
: acid-free paper)
      1. Holomorphic functions. 2. Fatou theorems. 3. Functions of
several complex variables. I. Title. II. Series: Progress in
mathematics (Boston, Mass.) ; vol. 147
QA331.D558 1997   95-30694
515--dc21   CIP

AMS Classification Codes: 31B25 and 32A40.

Printed on acid-free paper
© 1998 Birkhäuser

Birkhäuser

Copyright is not claimed for works of U.S. Government employees.
All rights reserved. No part of this publication may be reproduced, stored in a retrieval system, or transmitted, in any form or by any means, electronic, mechanical, photocopying, recording, or otherwise, without prior permission of the copyright owner.

Permission to photocopy for internal or personal use of specific clients is granted by Birkhäuser Boston for libraries and other users registered with the Copyright Clearance Center (CCC), provided that the base fee of $6.00 per copy, plus $0.20 per page is paid directly to CCC, 222 Rosewood Drive, Danvers, MA 01923, U.S.A. Special requests should be addressed directly to Birkhäuser Boston, 675 Massachusetts Avenue, Cambridge, MA 02139, U.S.A.

ISBN 0-8176-3976-4
ISBN 3-7643-3976-4
Typesetting by the author in LaTeX
Printed and bound by Hamilton Printing, Rensselaer, NY
Printed in the United States of America

9 8 7 6 5 4 3 2 1

Contents

Preface — vii

I Background — 1

1 Prelude — 3
  1.1 The Unit Disc — 3
  1.2 Spaces of Homogeneous Type — 16
  1.3 Euclidean Half-Spaces — 18
  1.4 Maximal Operators and Convergence — 24

2 Preliminary Results — 27
  2.1 Approach Regions — 27
  2.2 The Nagel–Stein Approach Regions — 43
  2.3 Goals, Problems and Results — 51

3 The Geometric Contexts — 55
  3.1 NTA Domains in $\mathbb{R}^n$ — 55
  3.2 Domains in $\mathbb{C}^n$ — 59
  3.3 Trees — 73

II Exotic Approach Regions — 85

4 Approach Regions for Trees — 87
  4.1 The Dyadic Tree — 87
  4.2 The General Tree — 88

5 Embedded Trees — 99
  5.1 The Unit Disc — 99
  5.2 Quasi-Dyadic Decompositions — 104
  5.3 The Maximal Decomposition of a Ball — 108

	5.4 Admissible Embeddings	110
6	**Applications**	**123**
	6.1 Euclidean Half-Spaces	124
	6.2 NTA Domains in $\mathbb{R}^n$	124
	6.3 Finite-Type Domains in $\mathbb{C}^2$	125
	6.4 Strongly Pseudoconvex Domains in $\mathbb{C}^n$	128

Notes   **129**

List of Figures   **131**

Guide to Notation   **133**

Bibliography   **135**

Index   **149**

Preface

A basic principle governing the boundary behaviour of holomorphic functions (and harmonic functions) is this: Under certain growth conditions, for almost every point in the boundary of the domain, these functions admit a boundary limit, if we approach the boundary point within *certain* approach regions. For example, for bounded harmonic functions in the open unit disc, the *natural* approach regions are nontangential triangles with one vertex in the boundary point, and entirely contained in the disc [Fat06]. In fact, these natural approach regions are optimal, in the sense that convergence will fail if we approach the boundary inside larger regions, having a higher order of contact with the boundary. The first theorem of this sort is due to J.E. Littlewood [Lit27], who proved that if we replace a nontangential region with the rotates of any fixed tangential curve, then convergence fails.

In 1984, A. Nagel and E.M. Stein proved that in Euclidean halfspaces (and the unit disc) there are in effect regions of convergence that are *not* nontangential: These *larger* approach regions contain tangential sequences (as opposed to tangential curves). The phenomenon discovered by Nagel and Stein indicates that the boundary behaviour of holomorphic functions (and harmonic functions), in theorems of Fatou type, is regulated by a second principle, which predicts the existence of regions of convergence that are *sequentially larger* than the natural ones. However, there are interesting (indeed typical) cases where the original technique, used to construct these larger approach regions, could not be applied, because of certain basic difficulties. We introduce a new method which shows that the Nagel–Stein phenomenon does indeed hold in great generality.

Recall that natural approach regions in theorems of Fatou type have been identified for the unit ball in $\mathbb{C}^n$ [Kor69], strictly pseudoconvex domains in $\mathbb{C}^n$ [Ste72], pseudoconvex domains of finite type in $\mathbb{C}^2$ [NSW81], [NSW85], NTA domains in $\mathbb{R}^n$ [JK82], symmetric spaces [Kor65], [Sjö86], [Ste83], certain negatively curved manifolds [AS85], [Anc87], certain Gromov-hyperbolic spaces [Anc90], e.g. trees [KP86].

The original construction of Nagel–Stein approach regions is based on the group of translations acting on Euclidean half-spaces: A certain *lacunary* approach region is constructed at a fixed point and then translated to other boundary points. More generally, if the boundary of the domain and the natural approach regions happen to "look alike at different points", then one can use similar constructions, whose precise application requires considerable care. Within the range of this general technique there fall the unit ball in $\mathbb{C}^n$ [Sue86], strictly pseudoconvex domains in $\mathbb{C}^n$ and smoothly bounded domain in $\mathbb{R}^n$ [AC92]; symmetric spaces and products of Euclidean half spaces [Sve95], [Sve96a],[Sve96b]. However, the boundary of pseudoconvex domains of finite type in $\mathbb{C}^2$ does not admit such a construct, since the shape and measure of the boundary balls do indeed change near weakly pseudoconvex points; a similar difficulty arises for NTA domains in $\mathbb{R}^n$.

Our result covers, in particular, the two notable cases that have so far escaped the reach of other methods, namely NTA domains in $\mathbb{R}^n$ and pseudoconvex domains of finite type in $\mathbb{C}^2$. In fact, we show that the Nagel–Stein phenomenon is independent of the presence of any (pseudo) group acting on the boundary. Our technique rests on a process of discretization, leading to the discrete setting of trees, for which the problem was solved in [ADBU96]. Observe that, without (pseudo-) group invariance, it is not sufficient to define pointwise the approach regions in a lacunary way, in order to obtain the desired bound on the relevant maximal operator.

We prove that the Nagel–Stein phenomenon holds under the following minimal assumptions, which recapture much that is common to different contexts: (1) the boundary of the domain is a *space of homogeneous type* without atoms; (2) the *shadow* projected on the boundary by the natural approach regions is uniformly comparable to a boundary ball; (3) to each boundary ball there corresponds a point in the domain, that is close to the ball and whose shadow is uniformly comparable to the ball itself.
Luleå, June 1997

Acknowledgments

I am deeply grateful to Steven Krantz for guiding my doctoral studies into a beautiful research area and to Elias Stein for a conversation we had at a crucial stage of this work. Special thanks to Massimo Picardello, Mitchell Taibleson and Guido Weiss for their teaching, encouragement and support. Vadim Kaimanovich, Juan Sueiro and Olof Svensson kindly provided comments and suggestions. Financial support and/or hospitality from CNR, EC-HCM, INdAM, Princeton University, Chalmers University of Technology, Luleå University and Rome "Tor Vergata" University are gratefully acknowledged.

Ai miei genitori

Fatou Type Theorems
Maximal Functions and Approach Regions

Part I

Background

Chapter 1

Prelude

1.1 The Unit Disc

Let U be an open subset of the complex plane $\mathbb{C}$. A smooth function $g : U \to \mathbb{R}$ is **harmonic** if the **Laplace equation**

$$\frac{\partial^2 g}{\partial x^2} + \frac{\partial^2 g}{\partial y^2} = 0$$

holds at every point of U. A smooth function $\Phi : U \to \mathbb{C}$ is **holomorphic** if the real and imaginary parts g and h of Φ satisfy the **Cauchy–Riemann equations**

$$\frac{\partial g}{\partial x} = \frac{\partial h}{\partial y}, \quad \frac{\partial g}{\partial y} = -\frac{\partial h}{\partial x}$$

in U; then g and h are harmonic, and h is said to be a **conjugate harmonic** function of g. It can be shown that Φ is holomorphic if and only if it is *representable in U by the power series* $\sum_0^\infty c_n(z - z_0)^n \equiv \Phi(z)$ for each $z_0 \in U$, $|z - z_0| < \inf_{\zeta \notin U} |\zeta - z_0|$.

In the **Dirichlet problem**, one is asked to determine a harmonic function with pre-assigned boundary values. The Dirichlet problem arises in mathematical physics and from the study of conformal mappings. In fact, B. Riemann [Rie], having solved the Dirichlet problem by means of a variational principle, deduced the existence of a biholomorphic map between the unit disc $D \stackrel{\text{def}}{=} \{z \in \mathbb{C} : |z| < 1\}$ and any simply connected domain strictly contained in $\mathbb{C}$. After K. Weierstrass [Wei] pointed out a flaw in the use of the variational principle, the need arose for another method to solve the Dirichlet problem. One of the results obtained by H.A. Schwarz [Sch72], a pupil of Weierstrass, in the study of this problem is the following basic theorem. Let $bD \stackrel{\text{def}}{=} \{z \in \mathbb{C} : |z| = 1\}$.

Theorem 1.1 [Sch72] *If $f : bD \to \mathbb{R}$ is continuous, then the Poisson integral*

$$Pf(z) \stackrel{def}{=} \frac{1}{2\pi} \int_0^{2\pi} \frac{1-|z|^2}{|e^{i\theta} - z|^2} f(e^{i\theta})\, d\theta \qquad (1.1)$$

is harmonic in D and

$$\lim_{\substack{z \in D \\ z \to e^{i\theta}}} Pf(z) = f(e^{i\theta}) \qquad (1.2)$$

for all $e^{i\theta} \in bD$. The convergence of z to $e^{i\theta}$ in (1.2) is unrestricted.

Proof. Since

$$\frac{1-|z|^2}{|z-w|^2} = \operatorname{Re}\left[\frac{w+z}{w-z}\right] \quad (z \in D,\ w \in bD), \qquad (1.3)$$

then the Poisson integral $Pf(z)$ is the real part of the function

$$\frac{1}{2\pi} \int_0^{2\pi} \frac{e^{i\theta}+z}{e^{i\theta}-z} f(e^{i\theta})\, d\theta, \qquad (1.4)$$

which is holomorphic in $z \in D$, being representable by power series in D. Thus $Pf(z)$ is harmonic in D. Moreover, (1.3) yields

$$\frac{1-|z|^2}{|z-w|^2} = \operatorname{Re}\left[\frac{1+z\bar{w}}{1-z\bar{w}}\right] = \left[1 + \sum_{k=1}^{\infty}(z\bar{w})^k + (\bar{z}w)^k\right] \qquad (1.5)$$

and therefore

$$\frac{1}{2\pi}\int_0^{2\pi} \frac{1-|z|^2}{|z-e^{i\theta}|^2}\, d\theta = 1 \quad (z \in D). \qquad (1.6)$$

Let $\theta_0 \in [0, 2\pi)$, and $\varepsilon > 0$. By (1.6), we may assume that $f(e^{i\theta_0}) = 0$. Observe that, as $z \to e^{i\theta_0}$, only the values of f near $e^{i\theta_0}$ are relevant to the behaviour of $Pf(z)$. In fact, for $n \in \mathbb{N}$, let I be the set of $w \in bD$ such that $|w - e^{i\theta_0}| < 1/n$, and write

$$Pf(z) = \frac{1}{2\pi}\int_I \frac{1-|z|^2}{|e^{i\theta}-z|^2} f(e^{i\theta})\, d\theta + \frac{1}{2\pi}\int_{bD\setminus I} \frac{1-|z|^2}{|e^{i\theta}-z|^2} f(e^{i\theta})\, d\theta\ .$$

Then, if $z \to e^{i\theta_0}$, $\frac{1}{2\pi}\frac{1-|z|^2}{|e^{i\theta}-z|^2} \lesssim (1-|z|)$ for each $e^{i\theta} \notin I$, and therefore $|\int_{bD\setminus I}| \lesssim (1-|z|)\int_0^{2\pi}|f|\, d\theta$. Now, choose n so large that $|f| < \varepsilon$ on I, and thus $|\int_I| < \varepsilon \frac{1}{2\pi}\int_0^{2\pi}\frac{1-|z|^2}{|e^{i\theta}-z|^2}\, d\theta = \varepsilon$, by (1.6). q.e.d.

1.1. THE UNIT DISC

The advent of the theory of Lebesgue integration [Leb02] gave new life to the study of Fourier series and made it possible to consider the Poisson integral (1.1) of an arbitrary (Lebesgue) integrable function $f \in L^1(bD)$. There are in fact intimate links between harmonic functions in D, Fourier series, and holomorphic functions in D. Observe that (1.5) implies the identity

$$Pf(re^{i\theta}) = \sum_{-\infty}^{\infty} r^{|n|} \hat{f}(n) e^{in\theta} = \sum_{0}^{\infty} r^n (a_n \cos n\theta + b_n \sin n\theta) \qquad (1.7)$$

between the Poisson integral $Pf(z) = Pf(re^{i\theta})$ of a (real-valued) function $f \in L^1(bD)$ and the Abel means of the Fourier series $S(f)$ of f:

$$S(f) \sim \sum_{-\infty}^{+\infty} \hat{f}(n) e^{in\theta} = \sum_{0}^{\infty} a_n \cos n\theta + b_n \sin n\theta, \qquad (1.8)$$

where

$$\hat{f}(n) = \frac{1}{2\pi} \int_0^{2\pi} f(e^{i\theta}) e^{-in\theta} d\theta \quad (n \in \mathbb{Z})$$

and

$$a_0 = \hat{f}(0), \quad a_n = 2\,\mathrm{Re}[\hat{f}(n)], \quad b_n = -2\,\mathrm{Im}[\hat{f}(n)] \quad n = 1, 2, \ldots.$$

Secondly, every harmonic function g in D admits a unique conjugate harmonic function h, so that $g + ih$ is holomorphic in D and $h(0) = 0$; in particular, since the Poisson integral Pf of a (real-valued) function $f \in L^1(bD)$ is the real part of the holomorphic function given in (1.4), and since

$$\mathrm{Im}\left[\frac{w+z}{w-z}\right] = \frac{2\,\mathrm{Im}[z\bar{w}]}{|w-z|^2} \quad (z \in D,\ w \in bD),$$

the conjugate harmonic function of Pf is given by

$$Qf(z) = \frac{1}{2\pi} \int_0^{2\pi} \frac{2\,\mathrm{Im}[ze^{-i\theta}]}{|e^{i\theta} - z|^2} f(e^{i\theta}) d\theta \quad (z \in D).$$

Since f is real-valued, $\hat{f}(-n) = \overline{\hat{f}(n)}$, and therefore

$$r^n \hat{f}(n) e^{in\theta} + r^n \hat{f}(-n) e^{-in\theta} = 2\,\mathrm{Re}[\hat{f}(n) r^n e^{in\theta}],$$

thus we deduce from (1.7) that $Qf(z) = Qf(re^{i\theta})$ is equal to

$$Qf(re^{i\theta}) = -i \sum_{-\infty}^{\infty} \mathrm{sign}(n) r^{|n|} \hat{f}(n) e^{in\theta} = \sum_{1}^{\infty} r^n (a_n \sin n\theta - b_n \cos n\theta)$$

where $\text{sign}(n) = \frac{n}{|n|}$ for $n \neq 0$ and $\text{sign}(0) = 0$. The partial sums

$$S_n(f)(\theta) = \sum_{k=-n}^{n} \hat{f}(k) e^{i\theta}$$

of the Fourier series (1.8) of a function are linked to a certain singular integral operator, the *conjugate function operator* $f \mapsto H(f)$

$$H(f)(e^{i\eta}) = -\frac{1}{2\pi} \int_{-\pi}^{\pi} \frac{f(\eta + \alpha)}{\tan \frac{\alpha}{2}} \, d\alpha$$

where the integral is interpreted as the limit

$$\lim_{\varepsilon \to 0} \int_{-\pi}^{-\varepsilon} + \int_{\varepsilon}^{+\pi} d\alpha$$

(if it exists). The conjugate function can also be recaptured by taking the boundary values of the harmonic conjugate of the Poisson extension of f

$$H(f)(e^{i\eta}) = \lim_{r \to 1} Qf(re^{i\eta}) \,.$$

The relation with the partial sums of the Fourier series of f is given as

$$S_n(f) = \sin_n H(f \cos_n) - \cos_n H(f \sin_n) + o(1) \qquad (1.9)$$

where $\sin_n(\theta) \stackrel{\text{def}}{=} \sin(n\theta)$, $\cos_n(\theta) \stackrel{\text{def}}{=} \cos(n\theta)$ and $o(1) \to 0$ as $n \to \infty$, uniformly in θ.

Lebesgue's theory found a receptive, fertile soil in the Moscow School of mathematics centered around D.F. Egorov and his pupil N.N. Luzin. Luzin used Theorem 1.4 [Luz13], [Luz16] to show that the singular integral operator $H(f)(e^{i\eta})$ converges for almost every η if $f \in L^2(\mathrm{b}D)$; he was convinced that the convergence of the singular integral defining $H(f)$, due to "a certain interference of positive and negative values" of the integrand, "should be considered as an actual cause for convergence of Fourier–Lebesgue series" of L^2 functions ([Luz16], quoted in [Dyn91, p. 197]; also in [Luz13]); his conjecture was confirmed by L. Carleson fifty years later, by the study of the maximal operator arising from (1.9). I.I. Privalov, Egorov's pupil, used Theorem 1.5, a result on the boundary behaviour of bounded holomorphic functions, to prove the convergence of $H(f)(e^{i\eta})$ for $f \in L^1(\mathrm{b}D)$, for almost every $e^{i\eta} \in \mathrm{b}D$.

If $f \in L^1(\mathrm{b}D)$ then the Poisson integral Pf is harmonic in D, and the proof of Theorem 1.1 shows that, if f is continuous at $e^{i\theta}$, then $Pf(z) \to f(e^{i\theta})$ as $z \to e^{i\theta}$; however, the limit in (1.2) does not exist for

1.1. THE UNIT DISC

every $e^{i\theta} \in bD$ as $z \to e^{i\theta}$. The following example, discussed in [Fat06, pp. 341–342], suggests that we still obtain a convergence result, provided we only consider certain points $w \in bD$, and the point z approaches w in a certain way. Consider a function $f \in L^1(bD)$ such that, for a certain point $\theta_0 \in [0, 2\pi)$, $\lim_{\varepsilon \to 0+} f(e^{i(\theta_0 \pm \varepsilon)}) = \pm \infty$. Denote by $|E|$ the Lebesgue measure of a (measurable) subset $E \subset bD$. For $z \in D$, let $d(z) \stackrel{\text{def}}{=} \inf_{w \in bD} |z - w| = 1 - |z|$ and consider

$$B_z \stackrel{\text{def}}{=} \{ w \in bD : |z - w| < 2d(z) \} . \tag{1.10}$$

Consider the tangential curve

$$\tau = \{ z \in D : d(z) = |z - e^{i\theta_0}|^2 \}$$

and the two sides τ^+ and τ^- of τ, namely

$$\tau^{\pm} = \{ z \in \tau : z = |z| e^{i(\theta_0 \pm \varepsilon)}, \text{ for some } \varepsilon > 0 \} .$$

Let $J_n = \{ w \in bD : 0 < |w - e^{i\theta_0}| < 1/n \}$ and

$$J_n^+ = \{ e^{i(\theta_0 + \varepsilon)} \in J_n : \varepsilon > 0 \} .$$

Choose $N \in \mathbb{N}$ so large that f is positive on J_N^+ and negative on $J \setminus J_N^+$. If $z \in \tau^+$ is close to $e^{i\theta_0}$, we write $Pf(z)$ as

$$Pf(z) = \int_{B_z} + \int_{bD \setminus J_N} + \int_{J_N^+ \setminus B_z} + \int_{J_N \setminus J_N^+} . \tag{1.11}$$

Then $\int_{J_N^+ \setminus B_z} > 0$. Observe that $P(z, \omega) \simeq \frac{d(z)}{|z-\omega|^2} \simeq \frac{1}{d(z)}$ for $w \in B_z$, and $|B_z| \simeq d(z)$, thus

$$\int_{B_z} P(z, e^{i\theta}) f(e^{i\theta}) d\theta \simeq \frac{1}{|B_z|} \int_{B_z} f(e^{i\theta}) d\theta \to +\infty \quad (\tau^+ \ni z \to e^{i\theta_0})$$

since $\lim_{\varepsilon \to 0+} f(e^{i(\theta_0 + \varepsilon)}) = +\infty$ and for each $n \in \mathbb{N}$, if $z \in \tau^+$ is close enough to $e^{i\theta_0}$, then $B_z \subset J_n^+$. Moreover, $|\int_{bD \setminus J_N}| \lesssim d(z) \int_0^{2\pi} |f|$ for $z \in \tau$. Finally, $|\int_{J_N \setminus J_N^+}| \lesssim \int_{J_N \setminus J_N^+} |f| \cdot 1 \leq \int_0^{2\pi} |f| < \infty$, for $z \in \tau^+$. It follows that the limit of $Pf(z)$, as $z \to e^{i\theta_0}$ and $z \in \tau^+$, is equal to $+\infty$. Similar reasoning shows that the limit along τ^- is equal to $-\infty$. Since the function Pf is continuous in D, it also follows that, as $z \to e^{i\theta_0}$, it will reach every limit, if we choose an appropriate path. Observe that $\int_{B_z} P(z, e^{i\theta}) f(e^{i\theta}) d\theta$ is the term in (1.11) that is responsible for the divergence of $Pf(z)$; for $f \geq 0$, this term is comparable to the average $\frac{1}{|B_z|}\int_{B_z} f(e^{i\theta})d\theta$ of f over B_z; since $z \in \tau^+$, this average *reads off* the discontinuity of f near $e^{i\theta_0}$. The following theorem, proved by H. Lebesgue [Leb04, pp. 120–125], provides a kind of continuity *on average* for functions in $L^1(bD)$. For $w \equiv e^{i\theta} \in bD$ and $0 < r$, consider the **arc** $B(w, r)$ in bD of radius r about w:

$$B(w, r) = \left\{ e^{i(\theta+\eta)} : |\eta| < r \right\} .$$

Theorem 1.2 (The Lebesgue Differentiation Theorem) *For each function $f \in L^1(bD)$, the set of points $w \in bD$ such that*

$$\lim_{r \to 0} \frac{1}{|B(w,r)|} \int_{B(w,r)} |f(e^{i\theta}) - f(w)|d\theta = 0 \qquad (1.12)$$

has full Lebesgue measure in bD.
If (1.12) holds, then

$$f(w) = \lim_{r \to 0} \frac{1}{|B(w,r)|} \int_{B(w,r)} f(e^{i\theta})d\theta . \qquad (1.13)$$

A point $w \in bD$ for which (1.12) holds is called a **Lebesgue point** of f. A family $\{B(w_s, r_s)\}_s$ of arcs in bD is said to converge **nontangentially** to $w \in bD$ if (i) $\lim_s r_s = 0$, and (ii) there is a constant $\alpha > 0$, independent of s, such that $w \in B(w_s, \alpha r_s)$. The Lebesgue Differentiation Theorem implies that

$$\lim_s \frac{1}{|B(w_s, r_s)|} \int_{B(w_s, r_s)} |f(e^{i\theta}) - f(w)|d\theta = 0$$

for each Lebesgue point w of f, and each family of arcs $\{B(w_s, r_s)\}_s$ converging nontangentially to w. Observe that, in the previous example, the family of arcs $\{B_z\}_{z \in \tau^+}$ *does not* converge nontangentially to $e^{i\theta_0}$, as $\tau^+ \ni z \to e^{i\theta_0}$. Thus, we are led to consider the limit of $Pf(z)$ as $z \in D$ tends to a *Lebesgue point w of f* while remaining *inside* the set

$$\{ z \in D : w \in B_z \} = \{ z \in D : |z - w| < 2d(z) \} .$$

More in general, for $\alpha > 0$, we are led to consider the **nontangential approach region** of width α at $e^{i\theta}$

$$\Gamma_\alpha(w) \stackrel{\text{def}}{=} \{ z \in D : |z - w| < (1 + \alpha)d(z) \} ,$$

as an effective approach region to w (observe that $\Gamma_\alpha(w)$ is contained in a triangle with vertex in w, interior to the disc D).

This heuristic is made precise in the following

1.1. THE UNIT DISC

Theorem 1.3 *For each $\alpha > 0$ there is a constant $c_\alpha > 0$, such that for all $f \in L^1(bD)$ and all $w \in bD$*

$$\sup_{z \in \Gamma_\alpha(w)} |Pf(z)| \leq c_\alpha \sup_{r > 0} \frac{1}{|B(w,r)|} \int_{B(w,r)} |f(e^{i\theta})| d\theta. \tag{1.14}$$

Proof. The argument given in Section 1.3 can be readily adapted to this situation. q.e.d.

P. Fatou [Fat06] used the Lebesgue differentiation theorem to prove the following theorems.

Theorem 1.4 [Fat06] *For any $f \in L^1(bD)$, the limit*

$$\lim_{\substack{z \in \Gamma_\alpha(w) \\ z \to w}} Pf(z) \tag{1.15}$$

exists and equals $f(w)$ for any $\alpha > 0$ and any Lebesgue point w of f.

Proof. Let w be a Lebesgue point of f. We may assume that $f(w) = 0$. Write $f = g + h$, where g is the product of f with the characteristic function of a suitably small arc around w. The Poisson integral of h is under control near w because h is zero near w. Then we use (1.14) to control the Poisson integral of g. q.e.d.

Theorem 1.5 [Fat06] *If Φ is a bounded holomorphic function in the unit disc, then the limit*

$$\lim_{\substack{z \in \Gamma_\alpha(w) \\ z \to w}} \Phi(z) \tag{1.16}$$

exists for almost every $w \in bD$.

In 1930 the paper [HL30] by Hardy and Littlewood appeared, in which (a version of) the Hardy–Littlewood *maximal function*

$$M'f(w) = \sup_{r > 0} \frac{1}{|B(w,r)|} \int_{B(w,r)} |f(e^{i\theta})| d\theta \tag{1.17}$$

was introduced, where $f \in L^1(bD)$ and $w \in bD$. This operator, and the very idea of taking the sup in order to study the limit of a sequence, turned out to be a powerful tool for the study of the convergence of a sequence of operators, in particular for the boundary behaviour of functions. We have already seen the basic role played by the Lebesgue Differentiation Theorem in the proof of Fatou's theorem. The main estimate for the maximal function is given by the following

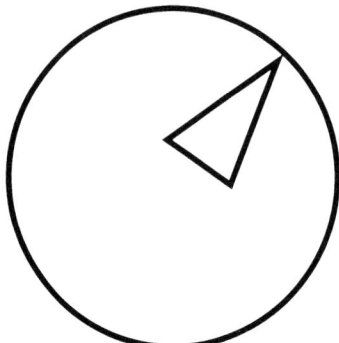

Figure 1.1: The nontangential approach region for the unit disc

Theorem 1.6 (The Maximal Theorem) *There is a constant $c > 0$ such that, for each $f \in L^1(bD)$, the inequality*

$$|\{w \in bD : M'f(w) > \lambda\}| \leq \frac{c}{\lambda} \int_0^{2\pi} |f(e^{i\theta})| d\theta$$

holds for any $\lambda > 0$.

(Cf. [DGS96].) The proof of Theorem 1.6 will be based on the corresponding result for the *dyadic maximal function*, introduced below.

For $n = 0, 1, 2, \ldots$, subsets of bD of the form

$$\{e^{i\theta} : \frac{2\pi}{2^n} k < \theta < \frac{2\pi}{2^n}(k+1)\}, \tag{1.18}$$

where k is an integer, $0 \leq k \leq 2^n - 1$, are called (open) **dyadic arcs** of generation n. Thus, $bD \setminus \{1\}$ is the (unique) dyadic arc of generation 0. Let T_n be the collection of all dyadic arcs of generation n, and let $T \stackrel{\text{def}}{=} \cup_{n=0}^\infty T_n$ be the (countable) collection of all dyadic arcs. On several occasions it is convenient to disregard the set

$$\mathcal{E} \stackrel{\text{def}}{=} \{e^{i2\pi k 2^{-n}} : n = 0, 1, 2, \ldots, k \text{ integer}, 0 \leq k \leq 2^n - 1\}$$

consisting of the end points of dyadic arcs; we may do so, since $\mathcal{E}$ is countable, thus a set of measure zero. In fact, for each $w \in bD \setminus \mathcal{E}$ and each $n = 0, 1, 2, \ldots$, there is a unique dyadic arc $T_n(w) \in T_n$ that contains w. If $I \in T_n$ and $n \geq 1$, the *predecessor* $\check{I}$ of I is the unique dyadic arc in T_{n-1} that contains I (thus, $bD \setminus \{1\}$ does not have a predecessor). Then

1.1. THE UNIT DISC

$|\tilde{I}| = 2|I|$. Observe that different dyadic arcs of the same generation are disjoint, and each dyadic arc I of generation n is contained in one and only one dyadic arc of generation k, for $0 \leq k < n$. Therefore, two dyadic arcs are either disjoint or one is contained in the other.

The union $\cup_{I \in \mathcal{A}} I \subset bD$ of all arcs belonging to a given collection $\mathcal{A} \subset T$ of dyadic arcs is denoted by $\cup \mathcal{A} \stackrel{\text{def}}{=} \cup_{I \in \mathcal{A}} I$. A collection $\mathcal{A} \subset T$ of dyadic arcs is called **disjointed** if its elements are pairwise disjoint.

Every non-empty collection $\mathcal{A} \subset T$ of dyadic arcs determines a unique *maximal* disjointed subcollection $\mathcal{A}_{\max}$ such that $\cup \mathcal{A} = \cup \mathcal{A}_{\max}$, as follows. If $\mathcal{A}$ contains $bD \setminus \{1\}$, we let $\mathcal{A}_{\max} = \{bD \setminus \{1\}\}$. If $\mathcal{A}$ does not contain $bD \setminus \{1\}$, then for each $I \in \mathcal{A}$ there is a largest dyadic arc $I_{\mathcal{A}}$ that contains I *and* belongs to $\mathcal{A}$; thus $I_{\mathcal{A}} \neq bD \setminus \{1\}$. Let

$$\mathcal{A}_{\max} \stackrel{\text{def}}{=} \{I_{\mathcal{A}} : I \in \mathcal{A}\}.$$

Then, a dyadic arc that properly contains an arc of $\mathcal{A}_{\max}$ does *not* belong to $\mathcal{A}$, $\mathcal{A}_{\max}$ is disjointed, and $\cup \mathcal{A} = \cup \mathcal{A}_{\max}$.

For $f \in L^1(bD)$, define the **dyadic maximal function of** f by

$$M_\triangle f(w) \stackrel{\text{def}}{=} \sup_{n \geq 0} \frac{1}{|T_n(w)|} \int_{T_n(w)} |f(e^{i\theta})| d\theta, \quad \text{if } w \in bD \setminus \mathcal{E}, \qquad (1.19)$$

and let $M_\triangle f(w) \stackrel{\text{def}}{=} 0$ if $w \in \mathcal{E}$.

Theorem 1.7 (The Dyadic Maximal Theorem)
If $f \in L^1(bD)$, then the inequality

$$|\{w \in bD : M_\triangle f(w) > \lambda\}| \leq \frac{1}{\lambda} \int_0^{2\pi} |f(e^{i\theta})| d\theta \qquad (1.20)$$

holds for each $\lambda > 0$.

Proof. Let $\mathcal{A} \stackrel{\text{def}}{=} \{I \in T : |I|^{-1} \int_I |f| d\theta > \lambda\}$. We may assume, without loss of generality, that $\mathcal{A}$ is not empty, and that it does not contain $bD \setminus \{1\}$ (for otherwise the conclusion is immediate). Observe that

$$\cup \mathcal{A} \setminus \mathcal{E} = \{w \in bD : M_\triangle f(w) > \lambda\}.$$

Since $\cup \mathcal{A} = \cup \mathcal{A}_{\max}$ and $\mathcal{A}_{\max}$ is disjointed, it follows that

$$|\{w \in bD : M_\triangle f(w) > \lambda\}| \leq \sum_{I \in \mathcal{A}_{\max}} |I|$$

$$\leq \sum_{I \in \mathcal{A}_{\max}} \frac{1}{\lambda} \int_I |f| d\theta \leq \frac{1}{\lambda} \int_0^{2\pi} |f| d\theta.$$

q.e.d.

An immediate consequence of Theorem 1.7 is the following

Theorem 1.8 (The Dyadic Differentiation Theorem)
If $f \in L^1(bD)$ then

$$f(w) = \lim_{n \to \infty} \frac{1}{|T_n(w)|} \int_{T_n(w)} f(e^{i\theta}) d\theta \tag{1.21}$$

holds for almost every $w \in bD$.

Proof. It suffices to prove that, for each positive integer j, the set

$$S_j = \left\{ w \in bD : \limsup_{n \to \infty} \left| \frac{1}{|T_n(w)|} \int_{T_n(w)} f(w) - f(e^{i\theta}) d\theta \right| > 1/j \right\}$$

has measure zero. Given $\varepsilon > 0$, let g be a continuous function on bD such that $f = g + h$, where the L^1 norm of h is smaller than ε. Thus, the set S_j is contained in the union of the three sets

$$\left\{ w \in bD : \limsup_n |T_n(w)|^{-1} \int_{T_n(w)} |h(e^{i\theta})| d\theta > \frac{1}{3j} \right\},$$

$$\left\{ w \in bD : \limsup_n |T_n(w)|^{-1} \left| \int_{T_n(w)} g(e^{i\theta}) - g(w) d\theta \right| > \frac{1}{3j} \right\}$$

and

$$\left\{ w \in bD : |h(w)| > \frac{1}{3j} \right\}.$$

The measure of the first set is bounded by $3j \int |h| d\theta < 3j\varepsilon$, by Theorem 1.7. Since g is continuous, the measure of the second set is zero. The measure of the third set is bounded by $3j \int |h| d\theta < 3j\varepsilon$. Thus, $|S_j| \leq 6j\varepsilon$. Therefore $|S_j| = 0$. **q.e.d.**

In order to relate the Hardy–Littlewood maximal function to the dyadic maximal function, we need the following

Lemma 1.9 *If $f \in L^1(bD)$ and $|I|^{-1} \int_I |f| \leq \lambda$ for each dyadic arc $I \in T$, then $|B|^{-1} \int_B |f| \leq \lambda$ for each arc $B \subset bD$.*

Proof. Let $B \subset bD$ be an arc in bD. Let $\mathcal{B} \stackrel{\text{def}}{=} \{I \in T : I \subset B\}$. Then $\mathcal{B}$ is nonempty. We may assume that $bD \setminus \{1\} \notin \mathcal{B}$. Observe that $B \setminus \mathcal{E} \subset \cup \mathcal{B} \subset B$, thus

$$B \setminus \mathcal{E} \subset \cup \mathcal{B}_{\max} \subset B$$

and therefore

$$|B|^{-1} \int_B |f| = \sum_n \frac{|J_n|}{|B|} |J_n|^{-1} \int_{J_n} |f| \leq \lambda \sum_n \frac{|J_n|}{|B|} = \lambda$$

1.1. THE UNIT DISC

where $\mathcal{B}_{\max} = \{J_n\}_n$. q.e.d.

Proof of Theorem 1.6. In view of Theorem 1.7, it suffices to show that there is a constant $c > 1$ such that

$$|\{w \in bD : M'f(w) > c\lambda\}| \leq 2|\{w \in bD : M_\triangle f(w) > \lambda\}| \quad (1.22)$$

for all $\lambda > 0$.

Consider again the collection $\mathcal{A} \stackrel{\text{def}}{=} \{I \in T : |I|^{-1} \int_I |f| d\theta > \lambda\}$. If $\mathcal{A}$ is empty, then Lemma 1.9 implies that the average of $|f|$ over each arc is bounded by λ, therefore $M'f(w) \leq \lambda$ for each $w \in bD$, hence (1.22) holds. If $\mathcal{A}$ contains $bD \setminus \{1\}$, then $M_\triangle f(w) > \lambda$ for each $w \in bD \setminus \{1\}$, hence (1.22) holds as well. Thus, we may assume that $\mathcal{A}$ is nonempty and does not contain $bD \setminus \{1\}$. If $I \in \mathcal{A}_{\max}$, then its predecessor $\tilde{I}$ (is defined and) does *not* belong to $\mathcal{A}$. Therefore $|\tilde{I}|^{-1} \int_{\tilde{I}} |f| d\theta \leq \lambda$. Thus

$$\lambda < |I|^{-1} \int_I |f| \leq \frac{|\tilde{I}|}{|I|} \frac{1}{|\tilde{I}|} \int_{\tilde{I}} |f| \leq 2\lambda \text{ for each } I \in \mathcal{A}_{\max}. \quad (1.23)$$

On the other hand, if $M_\triangle f(u) \leq \lambda$ and $u \notin \mathcal{E}$, then $\frac{1}{|T_n(u)|} \int_{T_n(u)} |f| \leq \lambda$ for each n, and thus Theorem 1.8 yields

$$|f(u)| \leq \lambda \text{ for almost every } u \notin \{M_\triangle f > \lambda\} \quad (1.24)$$

where $\{M_\triangle f > \lambda\} \stackrel{\text{def}}{=} \{w \in bD : M_\triangle f(w) > \lambda\}$.

If I is the dyadic arc given in (1.18), we denote by $2I$ the arc with the same *center* as I and twice the *radius*, i.e.,

$$2I \stackrel{\text{def}}{=} \left\{ e^{i\theta} : 2\pi 2^{-n} \frac{(2k-1)}{2} < \theta < 2\pi 2^{-n} \frac{(2k+3)}{2} \right\}.$$

Since $|2I| \leq 2|I|$ for each $I \in T$, and $\{M_\triangle f > \lambda\} = \mathcal{A}_{\max} \setminus \mathcal{E}$, then (1.22) will follow from

$$\{w \in bD : M'f(w) > c\lambda\} \subset \cup_{I \in \mathcal{A}_{\max}} 2I. \quad (1.25)$$

Let $w \notin \cup_{I \in \mathcal{A}_{\max}} 2I$, and let $B = B(w,r)$ be an open arc centered at w. Thus (1.24) yields

$$\frac{1}{|B|} \int_B |f| = \frac{1}{|B|} \int_{B \setminus \{M_\triangle f > \lambda\}} |f| + \frac{1}{|B|} \int_{B \cap \{M_\triangle f > \lambda\}} |f|$$

$$\leq \lambda + \frac{1}{|B|} \int_{B \cap \{M_\triangle f > \lambda\}}.$$

Recall again that $\{M_\triangle f > \lambda\}$ is equal to $\cup \mathcal{A}_{\max}$, modulo a null set, i.e. $\cup \mathcal{A}_{\max} \setminus \mathcal{E} = \{M_\triangle f > \lambda\}$. Now, if B has nonempty intersection with $I \in \mathcal{A}_{\max}$, then the radius of I is not greater than the radius of B (otherwise w would belong to $2I$); therefore $I \subset B(w, 3r)$. Let $\mathcal{B} \stackrel{\text{def}}{=} \{I \in \mathcal{A}_{\max} : I \cap B \neq \emptyset\}$. Then we get, by (1.23),

$$\begin{aligned}\frac{1}{|B|} \int_{B \cap \{M_\triangle f > \lambda\}} |f| &= \sum_{I \in \mathcal{B}} \frac{1}{|B|} \int_{I \cap B} |f| \leq \sum_{I \in \mathcal{B}} \frac{|I|}{|B|} \frac{1}{|I|} \int_I |f| \\ &\leq 2\lambda |B|^{-1} \sum_{I \in \mathcal{B}} |I| = 2\lambda |B|^{-1} |\cup \mathcal{B}| \\ &\leq 2\lambda |B|^{-1} |B(w, 3r)| \leq 6\lambda\end{aligned}$$

and therefore $\frac{1}{|B|} \int_B |f| \leq 7\lambda$ for each arc B centered at w. Thus $M'f(w) \leq 7\lambda$, and the proof of (1.25) is complete ($c = 7$). **q.e.d.**

Proof of Theorem 1.2. A simple argument shows that it suffices to show that (1.13) holds almost everywhere. This is achieved following the same reasoning as the proof of Theorem 1.8, using Theorem 1.6. **q.e.d.**

The conclusion of Theorem 1.5 also holds for holomorphic functions Φ in D subject to the growth condition

$$\sup_{0 < r < 1} \int_0^{2\pi} |\Phi(re^{i\theta})|^p d\theta < \infty, \tag{1.26}$$

$0 < p < \infty$, as well as for holomorphic functions subject to the (weaker) *Nevanlinna growth condition*

$$\sup_{0 < r < 1} \int_0^{2\pi} \log^+ |\Phi(re^{i\theta})| d\theta < \infty. \tag{1.27}$$

In 1932 [PZ32], R.E.A.C. Paley and A. Zygmund proved that the Nevanlinna condition cannot be weakened, in the following sense:

Theorem 1.10 *Given any non-negative measurable locally bounded function χ, defined on $[0, \infty)$ and such that $\chi(t) = o(t)$ as $t \to \infty$, there exists a holomorphic function Φ in D such that*

$$\sup_{0 < r < 1} \int_0^{2\pi} \chi(\log^+ |\Phi(re^{i\theta})|) d\theta < \infty$$

although $\lim_{r \to 1} \Phi(re^{i\theta})$ exists for almost no θ.

1.1. THE UNIT DISC

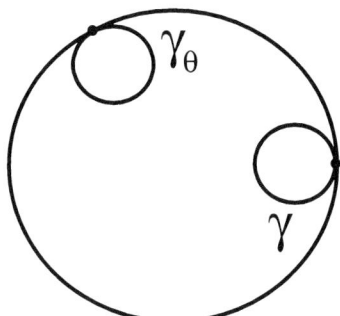

Figure 1.2: Littlewood's Theorem

The growth condition (1.26) defines the **Hardy space** of holomorphic functions $H^p(D)$. The conclusion of Theorem 1.5 applies to harmonic functions subject to the growth condition (1.26) *only* if we assume $1 \leq p \leq \infty$, as showed by Littlewood in 1931 [Lit31]. To wit, we have

Theorem 1.11 *There exists an harmonic function in D satisfying condition (1.26) for every positive $p < 1$, and without the property of possessing a radial limit for almost all θ.*

We quote a comment on the proof of Theorem 1.11, made by one of the editors of Littlewood's collected papers [Lit, p. 1045]:

> The examples are subtle, particularly [the one for Theorem 1.11]. A less recondite example of this fact would certainly be desirable, but to my knowledge no other example is known.

The extent by which it is necessary to avoid a tangential approach in the conclusion of Theorem 1.4 and Theorem 1.5 is described by the following theorem, proved by J. Littlewood [Lit27]. See Fig. 1.2.

Theorem 1.12 [Lit27] *Let $\gamma = \gamma_0 \subset D \cup \{1\}$ be a simple closed Jordan curve, having a common tangent with the circle at the point 1. Let γ_θ be the rotation of γ_0 by the angle θ. Then there exists a bounded holomorphic function $f(z)$ defined in D with the property that, for almost every $\theta \in [0, 2\pi]$, the limit of f along γ_θ does not exist.*

Other (stronger) versions of Theorem 1.12 were given by A.J. Lohwater and G. Piranian [LP] and Aikawa [Aik90] [Aik91].

In his proof, Littlewood used a nontrivial measure-theoretic result in Diophantine approximation proved by A.Ya. Khinchin, one of Luzin's pupils [Khi24, Satz II, p. 118]. In their book on number theory, Hardy and Wright refer to Khinchin's result as a "difficult" theorem [HW, Theorem 199, p. 169]. In 1949, A. Zygmund gave two different simpler proofs of Theorem 1.12 [Zyg49]. The first used complex analysis methods (Blaschke products). The second was entirely within the realm of real analysis, and it was the most enlightening; in retrospect we can read in it the elements of three later developments: the link between pointwise convergence and weak type estimates for maximal operators (Section 1.3 and Section 1.4); the Carleson tent condition (Section 2.1); the quasi-dyadic decomposition of a space of homogeneous type (Section 5.2).

The estimates for the Hardy–Littlewood maximal operator hold in great generality, in what are called *spaces of homogeneous type*, introduced in the next section. The boundary bD of the unit disc, with the Lebesgue linear measure and its natural metric, is an example of a space of homogeneous type.

1.2 Spaces of Homogeneous Type

A **quasi-metric space** (W, ρ) is a set W endowed with a **quasi-metric** ρ, a function $\rho : W \times W \to [0, \infty)$ such that there is a constant $A_0 \geq 1$ for which

$$\rho(w, u) = 0 \Leftrightarrow w = u$$

and

$$\rho(w, u) \leq A_0(\rho(w, v) + \rho(u, v))$$

for all $w, u, v \in W$. The **diameter** $\operatorname{diam}(A)$ of a subset A of W is the quantity $\operatorname{diam}(A) \stackrel{\text{def}}{=} \sup_{u,v \in A} \rho(u, v)$. In a quasi-metric space (W, ρ), consider the function

$$\mathbb{B} : W \times (0, \infty) \to 2^W$$

defined by

$$\mathbb{B}(w, r) \stackrel{\text{def}}{=} \{u \in W : \rho(w, u) < r\}, \quad w \in W, r > 0.$$

The function $\mathbb{B}$ is the **family of balls** associated to the quasi-metric space (W, ρ). Of special interest is the image $\mathbb{B}_W \subset 2^W$ of the map $\mathbb{B}$: Its elements are called **balls** in W. The map $\mathbb{B}$ need not be injective; w is

1.2. SPACES OF HOMOGENEOUS TYPE

called a **center** of $\mathbb{B}(w,r)$, and r a **radius** of $\mathbb{B}(w,r)$. A subset $A \subset W$ is **bounded** if it is contained in some ball.

If $c > 0$, we denote by $c \cdot \mathbb{B}$ the function $c \cdot \mathbb{B} : W \times (0, \infty) \to 2^W$ defined by

$$(c \cdot \mathbb{B})(w, r) \stackrel{\text{def}}{=} \{u \in W : \rho(w, u) < cr\}, \quad w \in W, r > 0.$$

A **space of homogeneous type** $W \equiv (W, \nu, \rho)$ is a topological space W equipped with a complete Borel measure ν and a quasi-metric ρ, which interact in the following way:

1. The family of concentric balls $\{\mathbb{B}(w,r)\}_{r>0}$ forms a neighborhood base of open sets at w, for each $w \in W$;

2. there is a constant A_1 such that

$$\nu(\mathbb{B}(w, 2r)) \leq A_1 \nu(\mathbb{B}(w, r))$$

for all $w \in W$ and $r > 0$ (the **doubling property** of the measure with respect to the quasi-metric);

3. $0 < \nu(\mathbb{B}(w, r)) < \infty$ for all $w \in W$ and $r > 0$.

We shall denote by $|A|$ the measure $\nu(A)$ of a measurable subset $A \subset W$. Given a function $f \in L^1(W, \nu)$, the **Hardy–Littlewood maximal function** Mf of f is the function

$$Mf(w) \stackrel{\text{def}}{=} \sup_{w \in B \in \mathbb{B}_W} \frac{1}{|B|} \int_B |f| \, d\nu, \quad w \in W. \tag{1.28}$$

The "radial" version of the Hardy–Littlewood maximal function is the function

$$M'f(w) \stackrel{\text{def}}{=} \sup_{r>0} \frac{1}{|\mathbb{B}(w,r)|} \int_{\mathbb{B}(w,r)} |f| \, d\nu, \quad w \in W. \tag{1.29}$$

A proof of the following basic result can be found in [CW71, Théorème 2.1, p. 71].

Theorem 1.13 *The inequality*

$$|\{w \in W : Mf(w) > \lambda\}| \lesssim \frac{1}{\lambda} \int_W |f| \, d\nu \tag{1.30}$$

holds for any $\lambda > 0$ and each $f \in L^1(W, \nu)$.

The inequality (1.30) is called an **estimate of weak type** $(1,1)$ for the operator $f \mapsto Mf$.

Metric spaces are not necessarily separable. The following result follows from [CW71, Lemme 1.1, p. 68].

Lemma 1.14 *All spaces of homogeneous type are separable and second countable.*

The following covering property of Whitney-type will be used in the sequel; for a proof, see [CW71, Théorème 1.3, p. 70] or [Ste93, Lemma 2, p. 15].

Lemma 1.15 *Let $O \subsetneq W$ be an open subset of a space of homogeneous type W. Then there is a sequence $\{\mathbb{B}(w_j, r_j)\}_{j \in \mathbb{N}}$ of balls such that*

1. *$O = \cup_j \mathbb{B}(w_j, r_j)$;*

2. *a point in O does not belong to more than m distinct balls in the family $\{\mathbb{B}(w_j, r_j)\}_{j \in \mathbb{N}}$;*

3. *each of the balls $\{h \cdot \mathbb{B}(w_j, r_j)\}_{j \in \mathbb{N}}$ intersects the complement of O.*

The constants m and h depend on W, but not on O.

1.3 The Euclidean Half Spaces $\mathbb{R}^n \times (0, \infty)$

A function u defined in an open subset of $\mathbb{R}^N$ is **harmonic** if it is twice continuously differentiable and the *Laplace equation* for $\mathbb{R}^N$

$$\sum_{j=1}^{N} \frac{\partial^2 u}{\partial x_j^2} = 0$$

holds at every point of the given open set.

The **Euclidean half-space**

$$\mathbb{R}_+^{n+1} \stackrel{\text{def}}{=} \mathbb{R}^n \times (0, \infty)$$

is a natural generalization of the half-plane $\{z \in \mathbb{C} : \text{Im}(z) > 0\}$[1], which is conformally equivalent to the unit disc. The boundary of $\mathbb{R}_+^{n+1}$ is identified with $\mathbb{R}^n \equiv \{(x, 0) : x \in \mathbb{R}^n\}$. It is a space of homogeneous type with respect to the Lebesgue measure dx and the Euclidean metric d. The Lebesgue measure of a subset E of $\mathbb{R}^n$ will be denoted by $|E|$; the

---
[1] $\text{Im}(z)$ is the imaginary part of z.

1.3. EUCLIDEAN HALF-SPACES

balls in $\mathbb{R}^n$ in the Euclidean metric d of $\mathbb{R}^n$ by $B(w, r)$. Points of $\mathbb{R}^{n+1}_+$ are denoted by
$$z = (w, t), \ w \in \mathbb{R}^n, \ t > 0.$$

We write
$$d(z) \stackrel{\text{def}}{=} \inf_{v \in \mathbb{R}^n} |z - v| = t \text{ if } z = (w, t)$$

where $|\cdot|$ is the Euclidean distance in $\mathbb{R}^{n+1}$, and
$$\pi(z) = w \text{ if } z = (w, t).$$

The **Poisson kernel** for $\mathbb{R}^{n+1}_+$ is given by

$$P(z; y) = P((w, t); v) = c_n \frac{t}{(t^2 + |w - v|^2)^{\frac{n+1}{2}}} \quad (1.31)$$

where $z = (w, t) \in \mathbb{R}^{n+1}_+$, $v \in \mathbb{R}^n$, $|w - v| = d(w, v)$ is the norm of $w - v$ in $\mathbb{R}^n$, and $\frac{1}{c_n}$ is half the surface area of the unit sphere in $\mathbb{R}^{n+1}$; see [SW71, pp. 7–9]. The function $P(z; v)$ is harmonic in $\mathbb{R}^{n+1}_+$ with respect to z. The **Poisson integral** of a function defined in $\mathbb{R}^n$ is given by

$$Pf(z) = \int_{\mathbb{R}^n} P(z; w) f(w) \, dw. \quad (1.32)$$

One has the following basic fact.

Theorem 1.16 *The Poisson integral of a continuous function on $\mathbb{R}^n$ with compact support solves the Dirichlet problem for $\mathbb{R}^{n+1}_+$.*

Proof. The proof is similar to that of Theorem 1.1, and will be omitted. See [SW71, pp. 47–48]. **q.e.d.**

In analogy with Theorem 1.4, we expect that, for any $f \in L^1(\mathbb{R}^n)$, the limit
$$\lim_{\substack{z \in \mathcal{G}(w) \\ z \to w}} Pf(z) \quad (1.33)$$

equals $f(w)$ for almost every $w \in \mathbb{R}^n$, if the region $\mathcal{G}(w) \subset \mathbb{R}^{n+1}_+$ has a conical shape. We now show that the shape of the region $\mathcal{G}(w)$ is forced on us by the nature of the Poisson kernel for $\mathbb{R}^{n+1}_+$; we therefore just assume, in this discussion, that a subset $\mathcal{G}(w) \subset \mathbb{R}^{n+1}_+$ is assigned to every point $w \in \mathbb{R}^n$, with the property that $\mathcal{G}(w)$ contains x in its closure, and try to derive its shape from the requirement that it gives the required convergence in (1.33).

In order to study the limit in (1.33), let us consider the quantity $\sup_{z\in\mathcal{G}(w)} Pf(z)$ or, better still, the operator

$$f \in L^1(\mathbb{R}^n) \mapsto \sup_{\mathcal{G}} P|f|$$

where

$$\sup_{\mathcal{G}} P|f| : \mathbb{R}^n \to [0,\infty]$$

is defined by

$$\sup_{\mathcal{G}} P|f|(w) \stackrel{\text{def}}{=} \sup_{z\in\mathcal{G}(w)} P|f|(z).$$

It is the *maximal operator associated* to the limit in (1.33). Thus, we take the pointwise least upper bound of the net for which we are trying to prove the existence of a limit. This idea was exploited for the first time in 1930 by Hardy and Littlewood in [HL30], where a version of the Hardy-Littlewood maximal function of (1.28) is introduced. According to A. Zygmund [Zyg76, p. 18], the idea of replacing the limit with a sup also appears in a 1909 paper by F. Jerosch and H. Weyl, but "unfortunately, that paper does not sufficiently exploit the brilliance of this idea".

Since we are trying to prove an almost everywhere result, we need only to be able to control the distribution function of

$$\sup_{z\in\mathcal{G}(w)} P|f|(z)$$

i.e. the quantity

$$|\{x \in \mathbb{R}^n : \sup_{z\in\mathcal{G}(w)} P|f|(z) > \lambda\}|$$

where λ is a positive number. More precisely, we need only the following weak type $(1,1)$ bound:

$$|\{w \in \mathbb{R}^n : \sup_{z\in\mathcal{G}(w)} P|f|(z) > \lambda\}| \lesssim \frac{1}{\lambda} \int_{\mathbb{R}^n} |f|\,d\nu \qquad (1.34)$$

for all $f \in L^1(\mathbb{R}^n)$ and any λ, in order to obtain the following:

Theorem 1.17 *For every $f \in L^1(\mathbb{R}^n)$, the limit*

$$\lim_{\substack{z\in\mathcal{G}(w) \\ z\to w}} Pf(z) \qquad (1.35)$$

exists and equals $f(w)$ for any $\alpha > 0$, and any $w \in \mathbb{R}^n$ outside a certain set of measure zero (which depends only on f).

1.3. EUCLIDEAN HALF-SPACES

Proof. Proceed as in the proof of Theorem 1.8, using (1.34) and Theorem 1.16. q.e.d.

Since we already know that the Hardy–Littlewood maximal operator satisfies the weak type estimate of (1.30), in order to obtain (1.34), it will be sufficient to select the region $\mathcal{G}(w)$ in such a way that the following distribution function inequality holds:

$$|\{w \in \mathbb{R}^n : \sup_{z \in \mathcal{G}(w)} P|f|(z) > \lambda\}| \lesssim |\{w \in \mathbb{R}^n : Mf(w) > \lambda\}| \quad (1.36)$$

for any $f \in L^1(\mathbb{R}^n)$ and all $\lambda > 0$. In particular, it suffices to select the region $\mathcal{G}(w)$ in such a way that the following pointwise estimate holds:

$$\sup_{z \in \mathcal{G}(w)} P|f|(z) \lesssim Mf(w) \quad (1.37)$$

for any $f \in L^1(\mathbb{R}^n)$ and all $\lambda > 0$. Thus, we are led to the problem of defining the region $\mathcal{G}(w) \subset \mathbb{R}^{n+1}_+$, for any $w \in \mathbb{R}^n$, in such a way that the pointwise inequality (1.37) holds.

Denote by 2^S the collection of all subsets of a given set S. Observe that any map $\mathcal{G} : \mathbb{R}^n \to 2^{\mathbb{R}^{n+1}_+}$ induces a map $\mathcal{G}^\downarrow : \mathbb{R}^{n+1}_+ \to 2^{\mathbb{R}^n}$, defined by

$$\mathcal{G}^\downarrow(z) = \{w \in \mathbb{R}^n : z \in \mathcal{G}(w)\}, \quad (1.38)$$

which we call the **shadow** of $\mathcal{G}$. Conversely, any map $\mathbf{g} : \mathbb{R}^{n+1}_+ \to 2^{\mathbb{R}^n}$ is the shadow of a unique map $\mathcal{G} : \mathbb{R}^n \to 2^{\mathbb{R}^{n+1}_+}$, define $\mathcal{G}(w) \stackrel{\text{def}}{=} \{z \in \mathbb{R}^{n+1}_+ : w \in \mathbf{g}(z)\}$. Therefore, it will be enough to describe the shadow map $\mathcal{G}^\downarrow$ in such a way that (1.37) holds. Observe that

$$P(z; v) \simeq \frac{d(z)}{|z - v|^{n+1}}$$

and therefore

$$P(z; v) \lesssim \frac{1}{(d(z))^n},$$

i.e., P is bounded by the reciprocal of the measure of a ball in $\mathbb{R}^n$ of radius $d(z)$: In view of the symmetry of P, it is natural to select the ball in $\mathbb{R}^n$ of center $\pi(z)$ and radius $d(z)$. Let

$$B(z) \stackrel{\text{def}}{=} B(\pi(z), d(z))$$

and write $P|f|(z)$ as

$$P|f|(z) = \int_{\mathbb{R}^n} P(z; v)|f(v)| \, dv$$

$$= \int_{B(z)} P(z;v)|f(v)|\, dv + \int_{\mathbb{R}^n \setminus B(z)} P(z;v)|f(v)|\, dv \stackrel{\text{def}}{=} I + II$$

$$\leq c \cdot \frac{1}{|B(z)|} \cdot \int_{B(z)} |f(v)|\, dv + II\ .$$

Since we seek a region $\mathcal{G}$ such that (1.37) holds, we need to know that

$$\frac{1}{|B(z)|} \cdot \int_{B(z)} |f(y)|\, dy$$

is bounded by $c \cdot Mf(w)$ for any $z \in \mathcal{G}(w)$, i.e., that $w \in B(z)$ for any $w \in \mathcal{G}^\downarrow(z)$. This means that $\mathcal{G}^\downarrow(z) \subset B(z)$. The easiest way to obtain this result is to *define* $\mathcal{G}^\downarrow(z)$ to be $B(z)$. This means that we define $\mathcal{G}(w)$ as the set $\{z : w \in \mathcal{G}^\downarrow(z)\} = \{z : w \in B(z)\} = \{z : w \in B(\pi(z), d(z))\} = \{z : |w - \pi(z)| < d(z)\}$, which is exactly a cone centered at the point w, with axis perpedicular to $\mathbb{R}^n$ and aperture of angle $\pi/4$, or width 1. That takes care of I.

It remains to estimate II. We break the integral into several pieces. Observe that, given $b > 0$ and $c > 1$, one has

$$\int_{c \cdot b \cdot B(z) \setminus b \cdot B(z)} |f(v)| P(z,v)\, dv$$

$$\leq \int_{cb \cdot B(z)} |f(v)| \frac{1}{b(b \cdot d(z))^n}\, dy \simeq \frac{1}{b} \frac{1}{|b \cdot B(z)|} \int_{cb \cdot B(z)} |f|\, dv$$

since $v \notin b \cdot B(z)$ implies that $|v - z| \geq b \cdot d(z)$. Now the doubling condition for $\mathbb{R}^n$ implies that $|cB(z)| \leq a_c |b \cdot B(z)|$ where the constant a_c only depends on c. Therefore,

$$\int_{c \cdot b \cdot B(z) \setminus b \cdot B(z)} |f(v)| P(z,v)\, dv \leq a_c \frac{1}{b} \frac{1}{|cb \cdot B(z)|} \int_{cb \cdot B(z)} |f|\, dv$$

$$\leq a_c \frac{1}{b} Mf(w)\ .$$

As a result, we break the term II into the sum of integrals over different rings with a fixed dilation factor $c = 2$: We get

$$II = \int_{\mathbb{R}^n \setminus B(z)} dy$$

$$= \int_{2B(z) \setminus B(z)} dv + \int_{4B(z) \setminus 2B(z)} dv + \int_{8B(z) \setminus 4B(z)} dv + \ldots$$

$$= \sum_{j=1}^{\infty} \int_{2^n B(z) \setminus 2^{n-1} B(z)} dv \leq \sum_{j=1}^{\infty} a_2 \cdot \frac{1}{2^{n-1}} Mf(w) \lesssim Mf(w)$$

1.3. EUCLIDEAN HALF-SPACES

obtaining the pointwise inequality of (1.37). Since dilation by a fixed constant does not affect these estimates, similar reasoning shows that (1.37) also holds for the cones $\mathcal{G}_\alpha \equiv \Gamma_\alpha$ given by

$$\mathcal{G}_\alpha(w_0) = \Gamma_\alpha(w_0) = \left\{ (w,t) \in \mathbb{R}^{n+1}_+ : |w - w_0| < \alpha \cdot t \right\}, \qquad (1.39)$$

where $w_0 \in \mathbb{R}^n$ and $\alpha > 0$.

This structure depends on the fact that (i) the boundary $\mathbb{R}^n$ of $\mathbb{R}^{n+1}_+$ is a space of homogeneous type and (ii) the agreement between the bounds for the relevant reproducing kernel (in this case, the Poisson kernel) and size of the measure of the balls in the boundary. It turns out that point (i) is a leitmotiv in the study of the boundary behaviour of functions, but there are cases in which point (ii) is not readily available, and the estimates will follow a different route, still leading to a bound in terms of Hardy–Littlewood maximal functions on the boundary. This technique of proving convergence results consists therefore of the following steps: (1) prove convergence for a dense class of functions; (2) prove that the boundary is a space of homogeneous type; (3) prove a bound for the maximal operator associated to the given approach region in terms of maximal operators associated to the boundary.

We have seen that the shape of the approach region $\mathcal{G}$, or, equivalently, its shadow $\mathcal{G}^\downarrow$, is forced on us by the requirement that the pointwise inequality of (1.37) holds, which led us to the set theoretic inclusion $\mathcal{G}^\downarrow(z) \subset B(z)$. This follows from the fact that we reduced the distribution function inequality of (1.34) to the pointwise inequality of (1.37). In Sections 2.1 and 2.2 we will see that if, instead, we only try to obtain the the distribution function inequality of (1.34), then we obtain a condition on the *measure* of the shadow $\mathcal{G}^\downarrow$ in terms of the measure of balls. This means that, if we want the family of approach regions $\mathcal{G}$ to be larger than the usual cones Γ_α, i.e., to contain tangential sequences, then we need the shadow $\mathcal{G}^\downarrow$ to be much larger than the corresponding balls, and this can only happen if the shadow $\mathcal{G}^\downarrow$ is not connected, i.e., if it has many holes. Then the region $\mathcal{G}$ will have many holes also.

Theorem 1.17 can be extended to harmonic functions with a growth condition similar to (1.26), for $1 \leq p \leq \infty$, which form the Hardy space $h^p(\mathbb{R}^{n+1}_+)$; cf. [SW71].

Littlewood's Theorem 1.12 has been extended to the Euclidean half space $\mathbb{R}^{n+1}_+$. A **curve** ending at $(0,0)$ is a continuous map $\gamma : (0,1) \to \mathbb{R}^{n+1}_+$ such that $\lim_{s \to 1} \gamma(s) = (0,0)$, the origin in $\mathbb{R}^{n+1}$; the curve γ is **tangential** to $\mathbb{R}^n$ if for each α there is an ε such that

$$(w,t) \in \gamma, 0 \leq t \leq \varepsilon \Rightarrow |w| \geq \alpha t.$$

Theorem 1.18 [Aik91] *Let $\gamma : (0,1) \subset \mathbb{R}_+^{n+1}$ be a a curve in $\mathbb{R}_+^{n+1}$ that ends at $(0,0)$ and is tangential to $\mathbb{R}^n$. Then there is a bounded harmonic function u in $\mathbb{R}_+^{n+1}$ such that for each $w \in \mathbb{R}^n$ the limit*

$$\lim_{t \to 0, (v,t) \in \gamma + w} u(v,t)$$

does not exist.

It is instructive to compare the proof of Theorem 1.18 with the real variable proof of Littlewood's Theorem 1.12 that was given by A. Zygmund in [Zyg49].

1.4 Maximal Operators and Convergence

A.N. Kolmogoroff, Luzin's pupil, proved that from the theorem of Privalov on the "existence" of the conjugate function $H(f)$ for $f \in L^1([0, 2\pi])$ (i.e., from a result on the boundary convergence of the harmonic conjugate to the Poisson extension of f) there follows a weak type estimate for the operator $f \mapsto H(f)$; cf. [Kol25]. In his proof, he proceeded by contraposition; from the assumption that the weak type estimate was false, he constructed a function in $L^1(\mathrm{b}D)$ for which the conjugate function diverged, contradicting Privalov's result. In the 1950s, Calderón proved a similar result, of a conditional nature, for the convergence of the Fourier series of L^2 functions. He showed that, assuming the convergence almost everywhere of the Fourier series S_n of L^2 functions, one could deduce that the operator

$$f \mapsto \sup_n |S_n(x)|$$

is of weak type $(2,2)$ [Zyg59, Theorem 1.22, p. 165].

That a general principle was in fact hidden behind these theorems was discovered by E.M. Stein in 1961 [Ste61]. He proved that a weak type estimate for the maximal function associated to a sequence of operators is a necessary condition for the existence of pointwise limits for this sequence of operators. The version given below is not the more general one proved in [Ste61], but it captures the main ingredients of the original statement.

Theorem 1.19 *Let $\{T_n\}_n$ be a sequence of bounded linear operators of $L^1(\mathrm{b}D)$ commuting with the rotations. Suppose that for every $f \in L^1(\mathrm{b}D)$, the limit*

$$\lim_{n \to \infty} T_n f(w)$$

1.4. MAXIMAL OPERATORS AND CONVERGENCE

exists for almost every $w \in bD$. Then the operator

$$f \mapsto \sup_n |T_n f(w)|$$

is of weak type $(1,1)$.

Extensions of Theorem 1.19 were given by [Saw66] and E.M. Nikisin (cf. [GCRdF85, Chapter 6]). We will see an application of Stein's theorem in Section 2.1, where we obtain a (weak) form of Littlewood's theorem.

The construction in [Kol25] is achieved by means of the family of mappings $z \mapsto z^n$, which is the prime example of a *mixing* family of measure-preserving transformation from bD into itself. The meaning of *mixing* is that it provides a way of getting *independent* or *almost independent* sets, thus enabling us to use a Borel–Cantelli argument, which ensures that a certain set has full measure. The argument is sketched in a footnote on p. 25 of [Kol25], and stated and proved in great generality, and greater precision, in [Saw66, Lemma 2, p. 165]. In his original paper [Ste61], E.M. Stein uses the Rademacher functions in order to get the independence (in [Ste61, Lemma 2, p. 147]; cf. also [Bur68] and [SZ93]). The Rademacher functions will also appear in Section 5.1, where they are used to obtain independence in the way the approach regions are defined at different points, thus ensuring that the resulting approach regions are *larger near the boundary* than the conical ones.

Chapter 2

Preliminary Results

2.1 Approach Regions

Let (W, ρ) be a quasi-metric space. A set D is a **space of approach to** (W, ρ), for the **approach function** $\tilde{\rho}: D \times W \to [0, \infty)$, if

1. for each $w \in W$ there is a sequence $\{z_n\}_n$ in D such that
$$\lim_{n \to \infty} \tilde{\rho}(z_n, w) = 0\,;$$

2. whenever
$$\lim_{n \to \infty} \rho(w_n, v_n) = 0 \text{ and } \lim_{n \to \infty} \tilde{\rho}(z_n, w_n) = 0$$
then
$$\lim_{n \to \infty} \tilde{\rho}(z_n, v_n) = 0\,;$$

3. whenever
$$\lim_{n \to \infty} \tilde{\rho}(z_n, w_n) = \lim_{n \to \infty} \tilde{\rho}(z_n, v_n) = 0$$
then
$$\lim_{n \to \infty} \rho(w_n, v_n) = 0,$$

where $\{z_n\}_n$ is a sequence in D and $\{w_n\}_n$, $\{v_n\}_n$ are sequences in W.

Example 2.1 *Let $\mathbb{B}_W$ be the collection of all balls in the quasi-metric space (W, ρ). For $B \in \mathbb{B}_W$ and $w \in W$, define*
$$\rho_W(B, w) = \sup_{u \in B} \rho(u, w)\,.$$

Then $\mathbb{B}_W$ is a space of approach to (W, d) for the approach function ρ_W.

Example 2.2 *If a subset D of a metric space has nonempty boundary* bD, *then D is a space of approach to* bD *for the induced metric.*

Let D be a space of approach to (W, ρ) for the approach function $\tilde{\rho}$. A sequence $\{z_n\}_n$ of points in D **converges to** $w \in W$ if
$$\lim_{n \to \infty} \tilde{\rho}(z_n, w) = 0 \,.$$
A subset $R \subset D$ is an **approach region to** $w \in W$ if there is a sequence in R that converges to w. A complex-valued function Φ defined on D **converges to** $\zeta \in \mathbb{C}$ **along** R **at** $w \in W$ if R is an approach region to $w \in W$ and for each $\varepsilon > 0$ there is a $\delta > 0$ such that $z \in R$ and $\tilde{\rho}(z, w) < \delta$ imply that $|\Phi(z) - \zeta| < \varepsilon$; we then write
$$\lim_{R \ni z \to w} \Phi(z) = \zeta \,.$$
If Φ converges to ζ along D at w then we simply write
$$\lim_{z \to w} \Phi(z) = \zeta \,.$$

An **approach family** $\{\mathcal{R}(w)\}_{w \in W}$ **for** (D, W) is given by the choice of a subset $\mathcal{R}(w)$ of D, for each point $w \in W$, with the property that for each $w \in W$, if the set $\mathcal{R}(w) \subset D$ is not empty, then it is an approach region to w. Thus, an approach family is a map $\mathcal{R} : W \to 2^D$ with the given properties. The **support** $\mathrm{supp}(\mathcal{R})$ of an approach family is the set of points $w \in W$ for which $\mathcal{R}(w)$ is nonempty.

Let Φ be a complex-valued function defined on D, $\mathcal{R}$ an approach family for (D, W), Q a subset of W, ϕ a function defined on Q. Then Φ **converges to** ϕ **along** $\mathcal{R}$ **on** $Q \subset W$ if Φ converges to $\phi(w)$ along $\mathcal{R}(w)$ at w, for each $w \in Q$.

Let $\{\mathcal{R}_n\}_{n \in \mathbb{N}}$ be a sequence of approach families for (D, W). Then the **union** $\cup_n \mathcal{R}_n$ is the approach family for (D, W) defined pointwise by
$$(\cup_n \mathcal{R}_n)(w) \stackrel{\mathrm{def}}{=} \cup_n \mathcal{R}_n(w) \,.$$
The **shadow** $\mathcal{R}^{\downarrow}$ of an approach family $\mathcal{R}$ for (D, W) is defined exactly as in (1.38). Thus $\mathcal{R}^{\downarrow}$ is the map $\mathcal{R}^{\downarrow} : D \to 2^W$ given by
$$\mathcal{R}^{\downarrow}(z) \stackrel{\mathrm{def}}{=} \{w \in W : z \in \mathcal{R}(w)\}$$
for $z \in D$. See Figure 2.1. In particular,[1]
$$z \in \mathcal{R}(w) \Leftrightarrow w \in \mathcal{R}^{\downarrow}(z)$$

---

[1]Observe that every map $\mathbf{r} : D \to 2^W$ is the shadow of precisely one approach family $\mathcal{R}$: Define $\mathcal{R}(w) \stackrel{\mathrm{def}}{=} \{z \in D : w \in \mathbf{r}(z)\}$). In fact, an approach family can be described as a subset R of $D \times W$. Then $(z, w) \in \mathsf{R} \Leftrightarrow z \in \mathcal{R}(w) \Leftrightarrow w \in \mathcal{R}^{\downarrow}(z)$.

2.1. APPROACH REGIONS

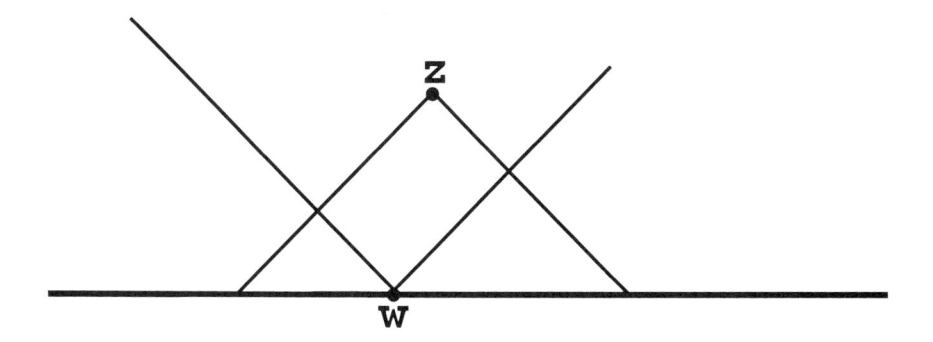

Figure 2.1: The relation between $\mathcal{L}$ and its shadow $\mathcal{L}^\downarrow$.

for $z \in D$ and $w \in W$.

The shadow $\mathcal{R}^\downarrow : D \to 2^W$ extends to a map $2^D \to 2^W$ which is also denoted $\mathcal{R}^\downarrow$, as follows: For $S \subset D$, we let

$$\mathcal{R}^\downarrow(S) \stackrel{\text{def}}{=} \cup_{z \in S} \mathcal{R}^\downarrow(z) .$$

In particular, the set

$$\mathcal{R}^\downarrow(S) = \{w \in W : S \cap L(w) \neq \emptyset\}$$

is the shadow of S projected by $\mathcal{R}$.

Let $\mathcal{R}$ be an approach family for (D, W), u a real-valued function defined on D, $\lambda > 0$, and $w \in W$. The following notation will be used[2]:

$$\sup_{\mathcal{R}} u(w) \stackrel{\text{def}}{=} \sup_{z \in \mathcal{R}(w)} u(z)$$

and

$$\{\sup_{\mathcal{R}} u > \lambda\} \stackrel{\text{def}}{=} \left\{ w \in W : \sup_{z \in \mathcal{R}(w)} u(z) > \lambda \right\} .$$

---

[2] $\sup \emptyset = 0$.

Proposition 2.3 *Let $\mathcal{G}$ be an approach family for (D,W). Then the following are equivalent:*

1. *$\mathcal{G}^{\downarrow}(z)$ is open for all $z \in D$;*

2. *the set*
$$\{\sup_{\mathcal{G}} u > \lambda\}$$
is open for each positive function u defined on D and each $\lambda > 0$.

Proof. $(1 \Rightarrow 2)$ Observe that
$$\{\sup_{\mathcal{G}} u > \lambda\} = \mathcal{G}^{\downarrow}(\{u > \lambda\}) = \bigcup_{z \in \{u > \lambda\}} \mathcal{G}^{\downarrow}(z)$$

is open, since it is the union of the open sets $\mathcal{G}^{\downarrow}(z)$. $(2 \Rightarrow 1)$ Fix $z \in D$. Let $u = 1$ at the point z and 0 elsewhere. Then $\{\sup_{\mathcal{G}} u > 1/2\} = \mathcal{G}^{\downarrow}(z)$, hence the conclusion. **q.e.d.**

If one of the two conditions of Proposition 2.3 holds for $\mathcal{G}$ then $\mathcal{G}$ is a **lower semi-continuous (l.s.c.) approach family**.

Let (W, ν, ρ) be a space of homogeneous type, and let D be a space of approach to (W, ρ) for the approach function $\tilde{\rho}$. An l.s.c. approach family $\mathcal{G}$ for (D, W) is called *natural* if the shadow $\mathcal{G}^{\downarrow}(z)$ of a point $z \in D$ close to the boundary is an open set in W of small diameter, close to z, and uniformly comparable to a ball: More precisely, $\mathcal{G}$ is a **natural approach family** if

1. if $w \in W$ and $z \in D$ then

    (a) $\mathcal{G}(w)$ is an approach region to w;

    (b) $\mathcal{G}^{\downarrow}(z)$ is open;

2. for every $z \in D$ there is a ball $\mathbb{B}(w_{\mathcal{G}}(z), r_{\mathcal{G}}(z))$ in W such that
$$k \cdot \mathbb{B}(w_{\mathcal{G}}(z), r_{\mathcal{G}}(z)) \subset \mathcal{G}^{\downarrow}(z) \subset k' \cdot \mathbb{B}(w_{\mathcal{G}}(z), r_{\mathcal{G}}(z))$$
    where k and k' are universal constants that do not depend on z;

3. if $\{z_n\}_n$ is a sequence in D converging to a point in W, then

2.1. APPROACH REGIONS

(a)
$$\lim_{n \to \infty} \operatorname{diam} \mathcal{G}^{\downarrow}(z_n) = 0$$

and

(b)
$$\lim_{n \to \infty} \sup_{u \in \mathcal{G}^{\downarrow}(z_n)} \tilde{\rho}(u, z_n) = 0 .$$

Example 2.4 Let (W, ν, ρ) be a space of homogeneous type. The collection $\mathbb{B}_W$ of all balls in W is a space of approach to W (Example 2.1). For $w \in W$, let $\mathfrak{B}_W(w) \subset \mathbb{B}_W$ be the collection of all balls in W containing w. Then $\mathfrak{B}_W$ is a natural approach family for $(\mathbb{B}_W, W)$. In fact, the shadow $\mathfrak{B}_W^*(B)$ of a point $B \in \mathbb{B}_W$ is precisely the ball $B \subset W$.

Example 2.5 Let D be the unit disc in $\mathbb{C}$. The boundary $\mathrm{b}D$ of D is a space of homogeneous type for the induced Euclidean metric and the Lebesgue measure, and D is a space of approach to $\mathrm{b}D$. For $w \in \mathrm{b}D$, let $\Gamma_1(w) \stackrel{\text{def}}{=} \{z \in D : |z - w| < 2(1 - |z|)\}$ be the nontangential approach region of width $\alpha = 1$. Then the shadow $\Gamma_1^{\downarrow}(z)$ is an open subset of $\mathrm{b}D$ that is uniformly comparable to a ball in $\mathrm{b}D$ in the induced metric. In fact, for $z \in D$, $z \neq 0$, let $w_\Gamma(z) \stackrel{\text{def}}{=} \frac{z}{|z|}, r_\Gamma(z) = 1 - |z|$, and let $w_\Gamma(0) \stackrel{\text{def}}{=} 1, r_\Gamma(0) = 1$. Then
$$\mathbb{B}(w_\Gamma(z), r_\Gamma(z)) \subset \Gamma_1^{\downarrow}(z) \subset 3\mathbb{B}(w_\Gamma(z), r_\Gamma(z)).$$

If $\mathcal{G}$ is a natural approach family for (D, W), then the **maximal operator** $M_\mathcal{G}$ **induced by** $\mathcal{G}$ is the operator $M_\mathcal{G}$ on $W = (W, \nu, \rho)$ defined in the following way:

$$M_\mathcal{G} f(w) \stackrel{\text{def}}{=} \sup_{z \in \mathcal{G}(w)} \frac{1}{|\mathcal{G}^{\downarrow}(z)|} \int_{\mathcal{G}^{\downarrow}(z)} |f| \, d\nu \qquad (2.1)$$

for $f \in L^1(W)$ and $w \in W$.

Example 2.6 Let (W, ν, ρ) be a space of homogeneous type. Consider $\mathfrak{B}_W$, the natural approach family for $(\mathbb{B}_W, W)$ defined in Example 2.4. Then the maximal operator $M_{\mathfrak{B}_W}$ induced by $\mathfrak{B}_W$ is precisely the Hardy-Littlewood maximal operator M defined in (1.28).

Proposition 2.7 If $\mathcal{G}$ is a natural approach family for (D, W), then
$$M_\mathcal{G} f(w) \underset{\sim}{<} M f(w)$$

for all $f \in L^1(W)$ and all $\lambda > 0$. In particular, $M_\mathcal{G}$ is of weak type $(1,1)$.

Proof. Let $z \in \mathcal{G}(w)$. Since $k \cdot B(w_\mathcal{G}(z), r_\mathcal{G}(z)) \subset \mathcal{G}^\downarrow(z)$, the doubling property implies that $|k' \cdot \mathbb{B}(w_\mathcal{G}(z), r_\mathcal{G}(z))| \lesssim |\mathbb{B}(w_\mathcal{G}(z), r_\mathcal{G}(z))| \lesssim |\mathcal{G}^\downarrow(z)|$, and therefore

$$\frac{1}{|\mathcal{G}^\downarrow(z)|} \int_{\mathcal{G}^\downarrow(z)} |f| d\nu \lesssim \frac{1}{|k' \cdot \mathbb{B}(w_\mathcal{G}(z), r_\mathcal{G}(z))|} \int_{k' \cdot \mathbb{B}(w_\mathcal{G}(z), r_\mathcal{G}(z))} |f| \, d\nu \, .$$

On the other hand, $\mathcal{G}^\downarrow(z) \subset k' \cdot \mathbb{B}(w_\mathcal{G}(z), r_\mathcal{G}(z))$ and $z \in \mathcal{G}(w)$ imply that $w \in k' \cdot \mathbb{B}(w_\mathcal{G}(z), r_\mathcal{G}(z))$, and therefore the last term in the previous inequality is bounded by $Mf(w)$. **q.e.d.**

Proposition 2.8 *If $\mathcal{G}$ is a natural approach family for (D, W) then*

$$\lim_{z \to w} \frac{1}{|\mathcal{G}^\downarrow(z)|} \int_{\mathcal{G}^\downarrow(z)} f \, d\nu = f(w)$$

for every $w \in W$ and each continuous function f on W.

Proof. Let $\delta > 0$. Let $\{z_n\}_n$ be a sequence in D converging to w. Let $w_n \in \mathcal{G}^\downarrow(z_n)$. Then $\tilde{\rho}(z_n, w_n) \to 0$ and $\tilde{\rho}(z_n, w) \to 0$ imply that $\rho(w_n, w) \to 0$. Since $\text{diam}\,(\mathcal{G}^\downarrow(z_n)) \to 0$ and $\rho(w_n, w) \to 0$, there is n_0 such that $\mathcal{G}^\downarrow(z_n) \subset \mathbb{B}(w, \delta)$ for each $n \geq n_0$, hence the conclusion. **q.e.d.**

If $A \subset W$ is a subset of W, not necessarily measurable, the **exterior measure** of A is the quantity

$$\nu_e(A) \stackrel{\text{def}}{=} \inf_{A \subset E} |E|$$

where E ranges over measurable subsets. Then ν_e coincides with the outer measure[3] induced by ν, as defined in [Fol84, p. 30].

If $\mathcal{G}$ is a natural approach family for (D, W) and $\mathcal{L}$ is another approach family for (D, W), then we seek conditions on $\mathcal{L}$ that imply the following **distributional inequality**

$$\nu_e(\{\sup_\mathcal{L} u > \lambda\}) \lesssim |\{\sup_\mathcal{G} u > \lambda\}| \qquad (2.2)$$

for all positive functions u defined in D and all $\lambda > 0$. Whenever the distributional inequality (2.2) holds, we say that the approach family $\mathcal{L}$ is **subordinate to $\mathcal{G}$**. The relevance of the distributional inequality (2.2) is given by the fact that, under fairly general conditions, it implies that *any* function converging *along* $\mathcal{G}$ will also converge *along* $\mathcal{L}$, almost everywhere, as in the following

---

[3] An **outer measure** on a space is a non-negative, monotonic, subadditive set function vanishing on the empty set.

2.1. APPROACH REGIONS

Theorem 2.9 *Assume that $\mathcal{G}$ is a natural approach family for (D,W), $\mathcal{L}$ is an approach family for (D,W) subordinate to $\mathcal{G}$, and $\text{supp}(\mathcal{L})$ has full measure[4] in W; moreover, assume that the measure ν of W is Radon. Then any function $\Phi: D \to \mathbb{C}$, which converges along $\mathcal{G}$ to a function $\phi: E \to \mathbb{C}$ on a subset $E \subset W$, also converges along $\mathcal{L}$ to ϕ on a subset of E of full measure in E.*

The proof of Theorem 2.9 relies on a *point-of-density argument*, which yields a certain geometric property of the *sawtooth regions* built from $\mathcal{G}$ and $\mathcal{L}$, as in A.P. Calderón's real-variable, n-dimensional version [Cal50] of the *local Fatou theorem*, first proved by I.I. Privalov [Pri] for the unit disc (by methods of complex analysis; cf. [Pri56]). A version in *group invariant product spaces* is given in [MPS90, Lemma 4.1, Theorem 4.1]; the present version follows in part [Wit, Lemma 5, Theorem 4]. See also [MPS89b], [MPS89a], [MS87].

We need a series of lemmata, where we use the following notation: For $w \in W$ and $r > 0$, let

$$\mathsf{B}_D(w,r) \stackrel{\text{def}}{=} \{z \in D : \tilde{\rho}(z,w) < r\}. \tag{2.3}$$

Lemma 2.10 *Assume the hypothesis of Theorem 2.9. Then*

$$\lim_{\mathcal{L}(w) \ni z \to w} \frac{1}{|\mathcal{G}^{\downarrow}(z)|} \int_{\mathcal{G}^{\downarrow}(z)} f \, d\nu = f(w)$$

for almost every $w \in W$ and each $f \in L^1(W)$.

Proof. Observe that Proposition 2.7 and 2.2 imply that

$$\nu_e(\{w \in W : \sup_{z \in \mathcal{L}(w)} \frac{1}{|\mathcal{G}^{\downarrow}(z)|} \int_{\mathcal{G}^{\downarrow}(z)} |h| \, d\nu > \lambda\}) \lesssim \frac{1}{\lambda} \int |h|$$

for each $h \in L^1(W)$, $\lambda > 0$. The set on the left-hand side may be nonmeasurable, yet we are showing that a certain set has zero measure, i.e., arbitrarily small exterior measure, so the usual argument applies. In fact, since ν is Radon, we can find a continuous function g on W such that $f = g + h$, where h has small L^1 norm. Then we can proceed as in the proof of Theorem 1.8, using Proposition 2.8. **q.e.d.**

The *sawtooth regions* appear in the following

---
[4] A set $E \subset S$ has **full measure in** S if $S \setminus E$ has (exterior) measure zero.

Lemma 2.11 *Assume the hypothesis of Theorem 2.9. Then for each bounded subset $E \subset W$ having positive exterior measure and each positive number r there is a subset $E' \subset E$ of full measure in E and a positive function $s(\cdot) > 0$ defined on E' such that*

$$\bigcup_{v \in E'} \mathcal{L}(v) \cap \mathsf{B}_D(v, s(v)) \subset \bigcup_{w \in E} \mathcal{G}(w) \cap \mathsf{B}_D(w, r).$$

The proof of Lemma 2.11 is based on the differentiation theorem *along* $\mathcal{L}$ given in Lemma 2.10, applied to the characteristic function of a possibly non-measurable set; as in [Ste70, p. 251], we use the following

Lemma 2.12 *Let E be a bounded subset, not necessarily measurable, of a space of homogeneous type W. Then there is a measurable subset $\tilde{E}$ containing E, such that $\nu_e(E) = |\tilde{E}|$. Moreover, for each measurable subset G of W, one has*

$$|\tilde{E} \cap G| = \nu_e(E \cap G).$$

Proof. Apply [Fol84, Proposition 1.13]. q.e.d.

Proof of Lemma 2.11. We may assume that the support of $\mathcal{L}$ is equal to W. Let $\tilde{E}$ be the measurable set given by Lemma 2.12, and apply Lemma 2.10 to the characteristic function of the set $\tilde{E}$. Then E admits a subset $E' \subset E$ of full measure in E such that

$$v \in E' \Rightarrow \lim_{\mathcal{L}(v) \ni z \to v} \frac{\nu_e(E \cap \mathcal{G}^{\downarrow}(z))}{|\mathcal{G}^{\downarrow}(z)|} = 1.$$

For $1/m < r_0$, $m \in \mathbb{N}$, let

$$E^m \stackrel{\text{def}}{=} \left\{ v \in E : L(v) \cap \mathsf{B}_D(v, 1/m) \subset \bigcup_{w \in E} \mathcal{G}(w) \cap \mathsf{B}_D(w, r) \right\}.$$

Then it suffices to show that

$$E' \subset \bigcup_{m > 1/r} E^m \subset E$$

i.e., that if $y \in E$ and $y \notin \bigcup_{m > 1/r} E^m$ then $y \notin E'$. Let $y \in E$ and assume that $y \notin \bigcup_{m > 1/r} E^m$. Thus for each $m > 1/r$ there is a point $z_m \in \mathcal{L}(y) \cap \mathsf{B}_D(y, 1/m)$ such that

$$z_m \notin \bigcup_{w \in E} \mathcal{G}(w) \cap \mathsf{B}_D(w, r). \qquad (2.4)$$

2.1. APPROACH REGIONS

In particular, the sequence $\{z_m\}_m$ converges to y. We claim that there exists

$$n_0 \in \mathbb{N} \text{ such that } m \geq n_0 \Rightarrow E \cap \mathcal{G}^{\downarrow}(z_m) = \emptyset. \qquad (2.5)$$

The claim implies that $y \notin E'$, since $z_m \in \mathcal{L}(y)$ and $z_m \to y$, thus concluding the proof. Next, we prove (2.5). Assuming that (2.5) fails, we will derive a contradiction. If (2.5) fails, then there are infinitely many values of m, say $\{m_j\}_j$ for which $E \cap \mathcal{G}^{\downarrow}(z_{m_j}) \neq \emptyset$. Thus there is a point $w_j \in E \cap \mathcal{G}^{\downarrow}(z_{m_j})$. Now, $z_{m_j} \to y$ and $w_j \in \mathcal{G}^{\downarrow}(z_{m_j})$ imply that $\tilde{\rho}(w_j, z_{m_j}) \to 0$ as $j \to \infty$, and therefore $z_{m_j} \in \mathsf{B}_D(w_j, r)$ if j is large enough. Thus $z_{m_j} \in \mathcal{G}(w_j) \cap \mathsf{B}_D(w_j, r)$ for $w_j \in E$, in contradiction with (2.4). **q.e.d.**

Lemma 2.11 immediately leads to a generalization of itself:

Lemma 2.13 *Assume the hypothesis of Theorem 2.9. Then for each bounded subset $E \subset W$ of positive exterior measure and each positive function $r(\cdot) > 0$ defined on E there is a subset $E' \subset E$ of full measure in E, together with a positive function $s(\cdot) > 0$ defined on E' such that*

$$\bigcup_{v \in E'} \mathcal{L}(v) \cap \mathsf{B}_D(v, s(v)) \subset \bigcup_{w \in E} \mathcal{G}(w) \cap \mathsf{B}_D(w, r(w)).$$

Proof. For $j \in \mathbb{N}$, let $E(j) \stackrel{\text{def}}{=} \{w \in E : r(w) > 1/j\}$. Apply Lemma 2.11 to the sets $E(j)$ having positive exterior measure. **q.e.d.**

Proof of Theorem 2.9. We may assume, without loss of generality, that the set E is bounded, the support of $\mathcal{L}$ is equal to W, and the function Φ is real-valued. We also assume that $\nu_e(E) > 0$, for otherwise there is nothing to prove. For $m \in \mathbb{Z}$ and $n \in \mathbb{N}$, let $E(m,n) \subset E$ be the set of all points $w \in E$ such that

1.
$$m/n \leq \phi(w) < (m+1)/n$$

   and

2. there exists $r(w) > 0$ such that

$$z \in \mathcal{G}(w) \cap \mathsf{B}_D(w, r(w)) \Rightarrow |\Phi(z) - \phi(w)| \leq 1/n.$$

Then the set $E(n) \stackrel{\text{def}}{=} \cup_m E(m,n)$ is equal to E. If $E(m,n)$ has positive exterior measure, then Lemma 2.13 yields a subset $E'(m,n) \subset E(m,n)$ of

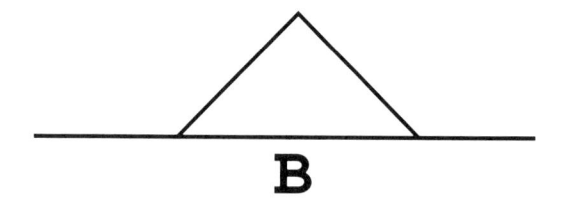

Figure 2.2: The Carleson tent for the Euclidean half-plane.

full measure in $E(m,n)$ and a positive function $s(\cdot)$ defined on $E'(m,n)$, such that

$$\bigcup_{v\in E'(m,n)} \mathcal{L}(v) \cap \mathsf{B}_D(v, s(v)) \subset \bigcup_{w\in E(m,n)} \mathcal{G}(w) \cap \mathsf{B}_D(w, r(w)).$$

Consider

$$\limsup_{\mathcal{L}(v)\ni z\to v} |\Phi(z) - \phi(v)| \stackrel{\text{def}}{=} \inf_{R>0} \sup_{z\in \mathsf{B}_D(v,R)} |\Phi(z) - \phi(v)|.$$

It follows that

$$v \in E'(m,n) \Rightarrow \limsup_{\mathcal{L}(v)\ni z\to v} |\Phi(z) - \phi(v)| \leq 2/n.$$

The set $E'(n) \stackrel{\text{def}}{=} \cup_m E'(m,n) \subset E$ has full measure in E, thus $\cap_n E'(n)$ also has full measure in E. Now, if $v \in \cap_n E'(n)$, then

$$\limsup_{\mathcal{L}(v)\ni z\to v} |\Phi(z) - \phi(v)| = 0.$$

q.e.d.

Having established, in Theorem 2.9, the relevance of the distributional inequality (2.2), we now show that there is indeed a condition on $\mathcal{L}$ and $\mathcal{G}$, the *tent condition*, which is equivalent to the distributional inequality (2.2) and does not depend on the function u, since it is formulated purely in terms of $\mathcal{L}$ and $\mathcal{G}$. We first introduce three useful notions. For a ball B in W, the subset of D given by

$$\triangle^{\mathcal{G}}(B) \stackrel{\text{def}}{=} \left\{ z \in D : \mathcal{G}^{\downarrow}(z) \subset B \right\} \tag{2.6}$$

is called the **Carleson tent over B along $\mathcal{G}$**. See Figure 2.2.

The proof of Theorem 2.9 shows a fact that will become clearer in the following theorem, namely that the distributional inequality (2.2) was

2.1. APPROACH REGIONS

used only for functions u on D of the form $u(z) \equiv \frac{1}{|\mathcal{G}^{\downarrow}(z)|} \int_{\mathcal{G}^{\downarrow}(z)} |f|\, d\nu$
where $f \in L^1(W)$. Given $f \in L^1(W)$, let

$$\mathcal{G}^{\div}(f)(z) \stackrel{\text{def}}{=} \frac{1}{|\mathcal{G}^{\downarrow}(z)|} \int_{\mathcal{G}^{\downarrow}(z)} |f|\, d\nu$$

for any $z \in D$, where $\mathcal{G}$ is a natural approach family for (D, W). The function $\mathcal{G}^{\div}(f)$ is the **average of** $|f|$ **along** $\mathcal{G}$. An approach family $\mathcal{L}$ for (D, W) induces an **outer measure on** D, denoted $\mathcal{L}^*$: If S is a subset of D, then $\mathcal{L}^*(S)$ is the exterior measure of the shadow of S, i.e.

$$\mathcal{L}^*(S) \stackrel{\text{def}}{=} \nu_e(\mathcal{L}^{\downarrow}(S)).$$

If $\mathcal{G}$ is a natural approach family then the approach family $\mathcal{L}$ satisfies the $\mathcal{G}$**-tent condition** if

$$\mathcal{L}^*(\Delta^{\mathcal{G}}(B)) \underset{\sim}{\leq} |B| \tag{2.7}$$

for all balls B in W.

In the following theorem we show that an approach family $\mathcal{L}$ is subordinate to $\mathcal{G}$ (i.e., the distributional inequality (2.2) holds for all positive functions u in W) if and only if the tent condition (2.7) holds for all balls B in W. The main idea of the proof, due to E.M. Stein [Ste70, p. 236], [FS71, pp. 113–114], is to consider the tents over the balls of a Whitney decomposition of the *open* set where $\sup_\mathcal{G} u$ is larger than λ; the use of this technique for the distributional inequality (2.2) involving two approach families was pointed out by J. Sueiro [Sue86], [Sue87], [Sue90, pp. 664–667] and by M. Andersson and H. Carlsson [AC92]. See also [Hör67].

Theorem 2.14 *If $\mathcal{G}$ is a natural approach family for (D, W) and $\mathcal{L}$ is another approach family for (D, W), then the following are equivalent:*

1. for all balls B in W

$$\mathcal{L}^*(\Delta^{\mathcal{G}}(B)) \underset{\sim}{\leq} |B|;$$

2. for all positive functions u on D and all $\lambda > 0$

$$\nu_e\{\sup_{\mathcal{L}} u > \lambda\} \underset{\sim}{\leq} |\{\sup_{\mathcal{G}} u > \lambda\}|;$$

3. for all $f \in L^1(W)$ and all $\lambda > 0$

$$\nu_e\{\sup_{\mathcal{L}} \mathcal{G}^{\div}(f) > \lambda\} \underset{\sim}{\leq} |\{\sup_{\mathcal{G}} \mathcal{G}^{\div}(f) > \lambda\}|;$$

4. for all $f \in L^1(W)$ and all $\lambda > 0$

$$\nu_e\{\sup_{\mathcal{L}} \mathcal{G}^{\div}(f) > \lambda\} \lesssim \frac{1}{\lambda}\int |f|d\nu.$$

Proof. We first show that (1) implies (2). Since the set $\{\sup_{\mathcal{G}} u > \lambda\}$ is open, we can apply the Whitney-type decomposition from Lemma 1.15, assuming, without loss of generality, that $\{\sup_{\mathcal{G}} u > \lambda\} \neq W$. This means that there are positive constants m and h, depending only on W, such that

1. the open set $\{\sup_{\mathcal{G}} u > \lambda\}$ can be written as the union of a countable family $\{\mathbb{B}(w_n, r_n)\}_n$ of balls;

2. each point of $\{\sup_{\mathcal{G}} u > \lambda\}$ belongs to not more than m distinct balls in the family $\{\mathbb{B}(w_n, r_n)\}_n$;

3. each ball $h \cdot \mathbb{B}(w_n, r_n)$ intersects the complement of $\{\sup_{\mathcal{G}} u > \lambda\}$.

Claim 2.15 *There is a universal constant c' such that*

$$\{u > \lambda\} \subset \bigcup_n \Delta^{\mathcal{G}}(c' \cdot \mathbb{B}(w_n, r_n))$$

for all n, where

$$\{u > \lambda\} \stackrel{\text{def}}{=} \{z \in D : u(z) > \lambda\}.$$

Assuming Claim 2.15, we conclude the proof as follows:

$$\begin{aligned}
\nu_e\{\sup_{\mathcal{L}} u > \lambda\} &= \nu_e\{w \in W : \mathcal{L}(w) \cap \{u > \lambda\} \neq \emptyset\} \\
&\leq \sum_n \nu_e(\mathcal{L}^{\downarrow}(\Delta^{\mathcal{G}}(c' \cdot \mathbb{B}(w_n, r_n)))) \lesssim \sum_n |c' \cdot \mathbb{B}(w_n, r_n)| \\
&\lesssim \sum_n |\mathbb{B}(w_n, r_n))| \lesssim |\cup_n \mathbb{B}(w_n, r_n)| = |\{\sup_{\mathcal{G}} u > \lambda\}|.
\end{aligned}$$

Proof of Claim 2.15. If $z \in \{u > \lambda\}$ then $\mathcal{G}^{\downarrow}(z) \subset \{\sup_{\mathcal{G}} u > \lambda\}$. Since $\mathcal{G}$ is a natural approach family, there is a ball $\mathbb{B}(w_{\mathcal{G}}(z), r_{\mathcal{G}}(z))$ such that

$$k \cdot \mathbb{B}(w_{\mathcal{G}}(z), r_{\mathcal{G}}(z)) \subset \mathcal{G}^{\downarrow}(z) \subset k' \cdot \mathbb{B}(w_{\mathcal{G}}(z), r_{\mathcal{G}}(z))$$

and, in particular, $w_{\mathcal{G}}(z) \in \mathbb{B}(w_{\mathcal{G}}(z), r_{\mathcal{G}}(z)) \subset \cup_n \mathbb{B}(w_n, r_n)$. Thus there is an integer $\bar{n}$ for which $w_{\mathcal{G}}(z) \in \mathbb{B}(w_{\bar{n}}, r_{\bar{n}})$. Therefore,

2.1. APPROACH REGIONS

1. $w_{\mathcal{G}}(z) \in \mathbb{B}(w_{\bar{n}}, r_{\bar{n}}) \subset h \cdot \mathbb{B}(w_{\bar{n}}, r_{\bar{n}}) \not\subset \{\sup_{\mathcal{G}} u > \lambda\}$, and

2. $k \cdot \mathbb{B}(w_{\mathcal{G}}(z), r_{\mathcal{G}}(z)) \subset \{\sup_{\mathcal{G}} u > \lambda\}$.

It follows that there is a constant c'', depending only on A_0, h and k, such that $r_{\mathcal{G}}(z) < c'' \cdot r_{\bar{n}}$. Therefore there is a constant c', depending only on A_0, k' and c'', such that $k' \cdot \mathbb{B}(w_{\mathcal{G}}(z), r_{\mathcal{G}}(z)) \subset c' \cdot \mathbb{B}(w_{\bar{n}}, r_{\bar{n}})$. Then $\mathcal{G}^{\downarrow}(z) \subset c' \cdot \mathbb{B}(w_{\bar{n}}, r_{\bar{n}})$, which means that $z \in \Delta^{\mathcal{G}}(c' \cdot \mathbb{B}(w_{\bar{n}}, r_{\bar{n}}))$. The proof of Claim 2.15 is thus completed.

It is clear that (2) implies (3). We deduce immediately from Proposition 2.7 that (3) implies (4). Finally, we show that (4) implies (1).

For a fixed ball B, consider the characteristic function χ_B of B. If $z \in \Delta^{\mathcal{G}}(B)$ then $\mathcal{G}^{\downarrow}(z) \subset B$ and therefore $\mathcal{G}^{\div}(\chi_B)(z) = 1$; in other words,

$$\Delta^{\mathcal{G}}(B) \subset \{\mathcal{G}^{\div}(\chi_B) > 1/2\}$$

therefore

$$\mathcal{L}^{\downarrow}(\Delta^{\mathcal{G}}(B)) \subset \mathcal{L}^{\downarrow}\{\mathcal{G}^{\div}(\chi_B) > 1/2\} = \{\sup_{\mathcal{L}} \mathcal{G}^{\div}(\chi_B) > 1/2\}$$

and thus

$$\nu_e(\mathcal{L}^{\downarrow}(\Delta^{\mathcal{G}}(B))) \leq |\{\sup_{\mathcal{L}} \mathcal{G}^{\div}(\chi_B) > 1/2\}| \lesssim \int |\chi_B| = |B|.$$

q.e.d.

In the construction of approach families, it is useful to consider the following notion. If $\mathcal{G}$ is a natural approach family for (D, W) and $\mathcal{L}$ is another approach family for (D, W) then the $\mathcal{G}$-**completion** $\mathcal{L}_{\mathcal{G}}$ **of** $\mathcal{L}$ is the approach family $\mathcal{L}_{\mathcal{G}}$ defined by

$$\mathcal{L}_{\mathcal{G}}(w) \stackrel{\text{def}}{=} \{z \in D : \mathcal{G}^{\downarrow}(z) \supset \mathcal{G}^{\downarrow}(\zeta) \text{ for some } \zeta \in \mathcal{L}(w)\} \quad (2.8)$$

for $w \in W$. Then $\mathcal{L}$ is $\mathcal{G}$-**complete** if $\mathcal{L} = \mathcal{L}_{\mathcal{G}}$.

An equivalent way to define the $\mathcal{G}$-completion of $\mathcal{L}$ is obtained by first observing that any approach family $\mathcal{G}$ for (D, W) extends as follows to a map

$$\mathcal{G}^{\uparrow} : D \to 2^D$$

where, for $\zeta \in D$, we let

$$\mathcal{G}^{\uparrow}(\zeta) \stackrel{\text{def}}{=} \{z \in D : \mathcal{G}^{\downarrow}(z) \supset \mathcal{G}^{\downarrow}(\zeta)\}.$$

Then

$$\mathcal{L}_{\mathcal{G}}(w) \equiv \bigcup_{\zeta \in \mathcal{L}(w)} \mathcal{G}^{\uparrow}(\zeta).$$

Example 2.16 *In the Euclidean half-space $\mathbb{R}^{n+1}_+$, the completion of an approach family $\mathcal{L}$ with respect to the cones Γ_α is obtained by adding to each point ζ of $\mathcal{L}(w)$ the closed cone $\Gamma_\alpha(\zeta)$, of aperture α, centered at the point, i.e., the cone*

$$\Gamma_\alpha(\zeta) \stackrel{\text{def}}{=} \left\{ z \in \mathbb{R}^{n+1}_+ : (\Gamma_\alpha)^\downarrow(z) \supset (\Gamma_\alpha)^\downarrow(\zeta) \right\}. \tag{2.9}$$

Thus, if $\zeta = (x,t) \in \mathbb{R}^n \times (0, \infty)$, then

$$\Gamma_\alpha(\zeta) = \left\{ (y,h) \in \mathbb{R}^{n+1}_+ : |y - x| \leq \alpha(h - t) \right\}.$$

In this context, an approach family is completed precisely when it satisfies the "cone condition" described in [NS84, p. 88]; see the following section.

Proposition 2.17 *If $\mathcal{G}$ is a natural approach family for (D,W), then an approach family $\mathcal{L}$ is subordinate to $\mathcal{G}$ if and only if the $\mathcal{G}$-completion $\mathcal{L}_\mathcal{G}$ is subordinate to $\mathcal{G}$.*

Proof. Observe that

$$\mathcal{L}^\downarrow(\Delta^\mathcal{G}(B)) = \mathcal{L}_\mathcal{G}{}^\downarrow(\Delta^\mathcal{G}(B))$$

and apply Theorem 2.14. q.e.d.

In practice, a natural approach family occurs as a member of a one-parameter collection $\{\mathcal{G}_\alpha\}_\alpha$ of natural approach families that are, in a sense, equivalent to each other. Let $\mathcal{G}$ be a natural approach family for (D,W) and $I \subset \mathbb{R}$ be an open interval. An **approach system (of dilates of $\mathcal{G}$)** is a collection $\{\mathcal{G}_\alpha\}_{\alpha \in I}$ of approach families $\mathcal{G}_\alpha$ for (D,W), where

1. $\mathcal{G}_\alpha$ is a natural approach family, for all $\alpha \in I$;
2. $\mathcal{G}_{\alpha_0} = \mathcal{G}$ for some $\alpha_0 \in I$;
3. $\mathcal{G}_\alpha(w) \subset \mathcal{G}_\beta(w)$ for all $w \in W$ and all $\alpha < \beta$ in I;
4. for each $z \in D$ there is a point $w_\mathcal{G}(z)$ and a positive number $r_\mathcal{G}(z)$ such that

$$k_\alpha \cdot \mathbb{B}(w_\mathcal{G}(z), r_\mathcal{G}(z)) \subset \mathcal{G}_\alpha{}^\downarrow(z) \subset k'_\alpha \cdot \mathbb{B}(w_\mathcal{G}(z), r_\mathcal{G}(z))$$

for all $z \in D$ and $\alpha \in I$, where the constants k_α, k'_α depend on $\alpha \in I$ but not on $z \in D$.

2.1. APPROACH REGIONS

The example below shows that the constants k_α, k'_α need not be uniform in α, but, in general, will diverge when α tends to the right-hand point of the interval I.

Example 2.18 *If $D = \mathbb{R}^{n+1}_+$ and $W = \mathbb{R}^n$, as in Section 1.3, then the family Γ_α of cones $\Gamma_\alpha(x) = \{z = (y,t) \in \mathbb{R}^{n+1}_+ : |y - x| < \alpha \cdot t\}$ is an approach system of dilates of $\Gamma \equiv \Gamma_1$, since its shadow $\Gamma^\downarrow_\alpha$ is the family*

$$\Gamma^\downarrow_\alpha(z) = \Gamma^\downarrow_\alpha(x,t) = \{y \in \mathbb{R}^n : |y - x| < \alpha \cdot t\} = \alpha \cdot \mathbb{B}(\pi(z), d(z))$$

obtained by projection on the boundary along "inverted cones".

Any two approach families $\mathcal{G}_\alpha$ and $\mathcal{G}_\beta$ belonging to a fixed approach system $\{\mathcal{G}_\alpha\}_{\alpha \in I}$ are *equivalent*, since the same family of balls

$$\{\mathbb{B}(w_\mathcal{G}(z), r_\mathcal{G}(z))\}_{z \in D}$$

can be used for each family of sets $\{(\mathcal{G}_\alpha)^\downarrow(z)\}_{z \in D}$; this equivalence can be seen in a more precise way, in the following

Proposition 2.19 *Let $\{\mathcal{G}_\alpha\}_{\alpha \in I}$ be an approach system for (D, W). Then for each $\alpha, \beta \in I$ there is a finite, positive constant $c_{\alpha,\beta}$ such that*

$$|\{\sup_{\mathcal{G}_\beta} u > \lambda\}| \leq c_{\alpha,\beta} |\{\sup_{\mathcal{G}_\alpha} u > \lambda\}|$$

for each positive function u defined on D and all $\lambda > 0$. In particular, an approach family is subordinate to $\mathcal{G}_\alpha$ if and only if it is subordinate to $\mathcal{G}_\beta$.

Proof. If $\beta \leq \alpha$ the conclusion follows from (3) in the definition of an approach system, so we assume that $\alpha < \beta$. We may assume that $|\{\sup_{\mathcal{G}_\alpha} u > \lambda\}|$ is finite, for otherwise there is nothing to prove. Then let $f \in L^1(W)$ be the characteristic function of the set $\{\sup_{\mathcal{G}_\alpha} u > \lambda\}$. Since the Hardy–Littlewood maximal operator is weak type $(1,1)$, by Theorem 1.13, then the conclusion follows (with a different constant $c_{\alpha,\beta}$) from the following claim:

$$\{\sup_{\mathcal{G}_\beta} u > \lambda\} \subset \{w \in W : Mf(w) > c_{\alpha,\beta}\}.$$

Assume then that $\sup_{\mathcal{G}_\beta} u(w_0) > \lambda$. Then there is a point $z_0 \in \mathcal{G}_\beta(w_0)$ such that $u(z_0) > \lambda$. Therefore

$$k_\alpha \cdot \mathbb{B}(w_\mathcal{G}(z_0), r_\mathcal{G}(z_0)) \subset \mathcal{G}^\downarrow_\alpha(z_0) \subset \{\sup_{\mathcal{G}_\alpha} u > \lambda\}.$$

Moreover,
$$w_0 \in \mathcal{G}_\beta^\downarrow(z_0) \subset k'_\beta \cdot \mathbb{B}(w_\mathcal{G}(z_0), r_\mathcal{G}(z_0))$$
and therefore the ball $k'_\beta \cdot \mathbb{B}(w_\mathcal{G}(z_0), r_\mathcal{G}(z_0))$ contains the point w_0. Therefore

$$\begin{aligned} Mf(w_0) &\geq \frac{|\{\sup_{\mathcal{G}_\alpha} u > \lambda\} \cap k'_\beta \cdot \mathbb{B}(w_\mathcal{G}(z_0), r_\mathcal{G}(z_0))|}{|k'_\beta \cdot \mathbb{B}(w_\mathcal{G}(z_0), r_\mathcal{G}(z_0))|} \\ &\geq \frac{|k_\alpha \cdot \mathbb{B}(w_\mathcal{G}(z_0), r_\mathcal{G}(z_0)) \cap k'_\beta \cdot \mathbb{B}(w_\mathcal{G}(z_0), r_\mathcal{G}(z_0))|}{|k'_\beta \cdot \mathbb{B}(w_\mathcal{G}(z_0), r_\mathcal{G}(z_0))|} \geq c_{\alpha,\beta}, \end{aligned}$$

where in the last inequality we have used the doubling property (and $c_{\alpha,\beta} = 1$ if it happens that $k'_\beta \leq k_\alpha$). **q.e.d.**

Corollary 2.20 *Let $D = \mathbb{R}_+^{n+1}$, $W = \mathbb{R}^n$, and Γ_α be the family of cones defined in Section 1.3. Then, Proposition 2.19 yields, for any positive function u on $\mathbb{R}_+^{n+1}$,*

$$|\{\sup_{\Gamma_\beta} u > \lambda\}| \leq c_{\alpha,\beta}|\{\sup_{\Gamma_\alpha} u > \lambda\}|$$

for all $\lambda > 0$.

From the point of view of convergence, the *equivalence* of the approach families $\mathcal{G}_\alpha$ belonging to an approach system $\{\mathcal{G}_\alpha\}_{\alpha \in I}$ is seen in the following

Theorem 2.21 *Let $\{\mathcal{G}_\alpha\}_{\alpha \in I}$ be an approach system for (D, W); moreover, assume that the measure of W is Radon. Then for $\alpha, \beta \in I$, any function $\Phi : D \to \mathbb{C}$, which converges along $\mathcal{G}_\beta$ to a function $\phi : E \to \mathbb{C}$ on a set $E \subset W$, also converges along $\mathcal{G}_\alpha$ to ϕ on a subset of E of full measure in E.*

Proof. Apply Proposition 2.19 and Theorem 2.9. **q.e.d.**

Let $\{\mathcal{G}_\alpha\}_{\alpha \in I}$ be an approach system for (D, W), and let $\mathcal{L}$ be an approach family for (D, W). We say that $\mathcal{L}$ is **exotic at the point** $w \in W$ with respect to $\{\mathcal{G}_\alpha\}_\alpha$, if $\mathcal{L}(w)$ contains a sequence $\{e_n(w)\}_{n \in \mathbb{N}}$ such that

1. $\lim_{n \to \infty} e_n(w) = w$, and

2. for each $\alpha \in I$ there are infinitely many $n \in \mathbb{N}$ such that $e_n(w) \notin \mathcal{G}_\alpha(w)$.

We say that $\mathcal{L}$ is **exotic on a subset** $E \subset W$, with respect to $\{\mathcal{G}_\alpha\}_\alpha$, if it is exotic at each point of E.

The interesting point of this definition[5] will be seen in the following section, where we present the construction, due to A. Nagel and E.M. Stein, of a *translation invariant* approach family $\mathcal{L}$ for the Euclidean half-space $\mathbb{R}^{n+1}_+$, which is *subordinate to the cones and exotic* at each point of the boundary, with respect to the cones. Thus, by Theorem 2.9, bounded harmonic functions in $\mathbb{R}^{n+1}_+$ and harmonic functions in the Hardy classes converge along $\mathcal{L}$ at almost every point in the boundary $\mathbb{R}^n$. This result should be compared with Theorem 1.12 and Theorem 1.18.

2.2 The Nagel–Stein Approach Regions

We now present the construction of a *translation invariant* approach family $\mathcal{L}$ for the Euclidean half-space $\mathbb{R}^{n+1}_+$, which is

1. subordinate to Γ_α for one (and therefore all) $\alpha \in I$;

2. exotic with respect to the approach system of cones.

In particular, $\mathcal{L}(w)$ contains a tangential sequence converging to w. This construction is due to A. Nagel and E.M. Stein, and is based on the possibility of defining the region at one point and then translating it at other points.

Let **L** be an open subset of $\mathbb{R}^{n+1}_+$ with the property that its closure contains the origin $(0,0)$. For each $h > 0$ let **L**$[h]$ be the **cross-section** of **L** at level h, i.e.,

$$\mathbf{L}[h] \stackrel{\text{def}}{=} \{w \in \mathbb{R}^n : (w,h) \in \mathbf{L}\}.$$

Let $\mathcal{L}$ be the approach family obtained from **L** by translations, i.e.,

$$\mathcal{L}(w) \stackrel{\text{def}}{=} \mathbf{L} + w,$$

where the addition acts on the first component of elements of **L**. An approach family obtained in this way is called **translation invariant**. Let $\mathcal{L}$ be a translation invariant approach family. The approach family $\mathcal{L}$ satisfies the **cone condition** (of aperture α) if it is complete with respect to the family of cones Γ_α, i.e., whenever **L** contains a point (w,r), it also contains the closed cone $\Gamma_\alpha(w,r)$ centered at (w,r), as defined in (2.9), i.e., the set

$$\Gamma_\alpha(w,r) \stackrel{\text{def}}{=} \{(w,r) + (x,y) : (x,y) \in \mathbb{R}^{n+1}_+, |x| \leq \alpha y\}.$$

---

[5] The Greek root *exo* means *outside*.

The following Theorem 2.22 is due to A. Nagel and E. M. Stein [NS84]. The proof we present here is adapted from J. Sueiro [Sue86].

Theorem 2.22 *If* $\mathbf{L}$ *satisfies the cone condition, then the following conditions are equivalent:*

1. $|\mathbf{L}[h]| \lesssim h^n$ *for all* $h > 0$—*this is the* **cross section condition for** $\mathbf{L}$;

2. *let* $\mathcal{L}(w) \stackrel{\text{def}}{=} w + \mathbf{L}$; *then* $\mathcal{L}$ *is subordinate to the approach family of cones (say, of aperture one, thus of any aperture).*

Proof. Observe that h^n is (comparable to) the measure of a ball of radius h in $\mathbb{R}^n$. In view of Theorem 2.14 it is enough to show that the tent condition for $\mathcal{L}$ is equivalent to the cross-section condition for $\mathbf{L}$. We use the notation

$$S_1 \pm S_2 \stackrel{\text{def}}{=} \{x \pm y : x \in S_1, y \in S_2\}$$

and

$$-S_1 \stackrel{\text{def}}{=} \{-x : x \in S_1\}$$

where S_1 and S_2 are subsets of $\mathbb{R}^n$. Denote by r_B the radius of a ball B in $\mathbb{R}^n$. Let 0 be the origin in $\mathbb{R}^n$. Recall that for $z \in \mathbb{R}^{n+1}_+$, $d(z)$ is the distance to the boundary $\mathbb{R}^n$ and $\pi(z)$ the projection on the boundary. The proof is reduced to the verification of the following three facts:

$$\mathcal{L}^\downarrow(\triangle^\Gamma(B)) \subset B - \mathbf{L}[r_B] \subset \mathcal{L}^\downarrow(\triangle^\Gamma(2 \cdot B)); \tag{2.10}$$

$$|B - \mathbf{L}[r_B]| = |\mathbf{L}[r_B] + B(0, r_B)|; \tag{2.11}$$

$$\mathbf{L}[r_B] \subset \mathbf{L}[r_B] + B(0, r_B) \subset \mathbf{L}[3r_B]. \tag{2.12}$$

In fact, because of the doubling property for $\mathbb{R}^n$, given (2.10), (2.11) and (2.12), a bound for $\mathcal{L}^\downarrow(\triangle^\Gamma(B))$ in terms of $|B|$ implies a bound for $\mathbf{L}[r_B]$ in terms of $|B|$, and conversely. Let us prove (2.10). If $x \in \mathcal{L}^\downarrow(\triangle^\Gamma(B))$, where B is a ball in $\mathbb{R}^n$, then there is a point $z \in \mathcal{L}(x)$ such that $z \in \triangle^\Gamma(B)$. In particular, $d(z) < r_B$ and $\pi(z) \in B$. Now, $z = (\pi(z), d(z)) = (x+\eta, d(z)) = x + (\eta, d(z))$, where evidently $\eta = \pi(z) - x$ and $(\eta, d(z)) \in \mathbf{L}$. Since $d(z) < r_B$, the cone condition implies that $(\eta, r_B) \in \mathbf{L}$. Therefore $x = \pi(z) - \eta$ where $\pi(z) \in B$ and $\eta \in \mathbf{L}[r_B]$, since $(\eta, r_B) \in \mathbf{L}$. This proves that $x \in B - \mathbf{L}[r_B]$, i.e., $\mathcal{L}^\downarrow(\triangle^\Gamma(B)) \subset B - \mathbf{L}[r_B]$. The proof of the fact that $B - \mathbf{L}[r_B] \subset \mathcal{L}^\downarrow(\triangle^\Gamma(2 \cdot B))$ is similar, and is left to the reader.

2.2. THE NAGEL–STEIN APPROACH REGIONS

The proof of (2.11) depends on the following invariance properties of the Euclidean metric and the Lebesgue measure: $|S| = |-S|$, $|x + S| = |S|$ and $-B(0, r) = B(0, r)$. In fact, if $B = B(x, r)$ then

$$\begin{aligned} |B - \mathbf{L}[r_B]| &= |x + B(0, r) - \mathbf{L}[r_B]| \\ &= |B(0, r) - \mathbf{L}[r_B]| = |\mathbf{L}[r_B] - B(0, r)| = |\mathbf{L}[r_B] + B(0, r)| \end{aligned}$$

We finally prove (2.12). Since it is clear that $\mathbf{L}[r_B] \subset \mathbf{L}[r_B] + B(0, r_B)$, we need only show that $\mathbf{L}[r_B] + B(0, r_B) \subset \mathbf{L}[3r_B]$. If $w \in \mathbf{L}[r_B] + B(0, r_B)$ then there is a point $y \in \mathbf{L}[r_B]$ such that $w \in B(y, r_B)$. Now, $(y, r_B) \in \mathbf{L}$ and $|w - y| < r_B$ imply that $(w, 3 \cdot r_B) \in \mathbf{L}$, because $\mathbf{L}$ satisfies the cross-section condition. This means that $w \in \mathbf{L}[3 \cdot r_B]$.   q.e.d.

If $\mathcal{L}$ is any approach family for $(\mathbb{R}^n, \mathbb{R}^{n+1}_+)$, not necessarily invariant under translation, then we may consider the cross-section of $\mathcal{L}$ at the point w at height h, i.e., the set

$$\mathcal{L}(w)[h] \stackrel{\text{def}}{=} \{ v \in \mathbb{R}^n : (v, h) \in \mathcal{L}(w) \},$$

and consider the corresponding cross-section condition

$$|\mathcal{L}(w)[h]| \lesssim h^n, \text{ for all } w \in \mathbb{R}^n, h > 0. \tag{2.13}$$

Examples show that, without translation invariance, a cross-section condition does not yield the distributional inequality. We shall return to this issue after the proof of the following

Proposition 2.23 *There is a subset* $\mathbf{L}$ *of* $\mathbb{R}^{n+1}_+$, *whose closure contains the origin of* $\mathbb{R}^n$, *with the following properties:*

1. $\mathbf{L}$ *satisfies the cone condition;*

2. $|\mathbf{L}[h]| \lesssim h^n$ *for each* $h > 0$;

3. *the set* $\mathbf{L}$ *contains a sequence* $\{z_j\}_{j \in \mathbb{N}}$ *converging to the origin* 0 *of* $\mathbb{R}^n$, *such that for each* α *there is a subsequence* $\{z_{j_k}\}_k$ *of* $\{z_j\}_{j \in \mathbb{N}}$ *which is not contained in* $\Gamma_\alpha(0)$.

Proof. Fix a curve $\mathcal{C}$ in $\mathbb{R}^{n+1}_+$ that is tangential to the boundary $\mathbb{R}^n$ at the origin. Assume for simplicity that the curve is given by a continuous increasing function $\phi : (0, 1) \to (0, 1)$, such that $\lim_{x \to 0} \frac{\phi(x)}{x} = 0$, and that the curve $\mathcal{C}$ is contained in $\{(x, 0, 0, \ldots, y) : x \in \mathbb{R}, y > 0\} \subset \mathbb{R}^{n+1}_+$. Thus

$$\mathcal{C} = \{(x, 0, 0, \ldots, 0, \phi(x)) : x \in (0, 1)\} \subset \mathbb{R}^{n+1}_+$$

(where 0 is repeated exactly $n-1$ times). In what follows, we identify a point (x,y) of the half-plane $\mathbb{R} \times (0,\infty)$, with the corresponding point $(x,0,0,\ldots,y)$ in $\mathbb{R}^{n+1}_+$. The set $\mathbf{L}$ is obtained as the completion, with respect to Γ_1, of a sequence $\Sigma = \{z_j\}_{j \in \mathbb{N}} = \{(x_j, y_j)\}_{n \in \mathbb{N}}$ of points belonging to the curve $\mathcal{C}$. This means that

$$\mathbf{L} \equiv \Sigma_{\Gamma_1} \equiv \bigcup_{z \in \Sigma} \Gamma_1(z)$$

(see Definition 2.9).

All we have to do is to select the sequence

$$\{z_j\}_{j \in \mathbb{N}} = \{(x_j, y_j)\}_{j \in \mathbb{N}} = \{(x_j, \phi(x_j))\}_{j \in \mathbb{N}} \in \mathcal{C}$$

in such a way that

1. the sequence itself is tangential, i.e.,

$$\lim_{j \to \infty} \frac{\phi(x_j)}{x_j} = \lim_{j \to \infty} \frac{y_j}{x_j} = 0, \text{ and}$$

2. the cross-section condition for $\mathbf{L}$ holds.

The following claim is contained in [NS84, Lemma 9].

Claim 2.24 *If the sequence $\Sigma \equiv \{z_j\}_{j \in \mathbb{N}} \equiv \{(x_j, y_j)\}_{j \in \mathbb{N}} \subset \mathbb{R}^{n+1}_+$ has the property that the sequence $\{(x_{j+1}, y_j) \in \mathbb{R}^{n+1}_+\}_{n \in \mathbb{N}}$ belongs to the cone $\Gamma_\beta(0,0) \subset \mathbb{R}^{n+1}_+$, i.e., that*

$$|x_{j+1}| \leq \beta y_j \text{ for all } j \in \mathbb{N},$$

and the sequence $\{y_j\}_{j \in \mathbb{N}}$ is decreasing and converges to 0 as $j \to \infty$, then the cross-section condition

$$|\mathbf{L}[h]| \leq c \cdot h^n$$

holds for

$$\mathbf{L} \stackrel{\text{def}}{=} \Sigma_{\Gamma_1} \equiv \bigcup_{j \in \mathbb{N}} \Gamma_1(z_j).$$

Moreover, the constant c depends only on β and the dimension n.

2.2. THE NAGEL–STEIN APPROACH REGIONS

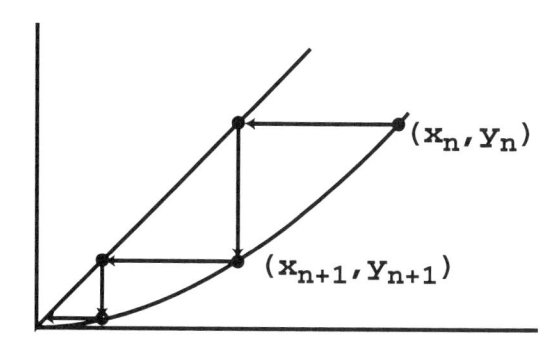

Figure 2.3: The construction of the sequence $\{(x_n, y_n)\}_{n \in \mathbb{N}}$.

Proof of Claim 2.24. Fix a positive h. There is a first index j_0 such that $y_{j_0} = \phi(x_{j_0})$ is less than h. Then the cross-section $\mathbf{L}[h]$ is the union of two parts: the cross-section for $\Gamma_1(z_{j_0})[h]$, and the one for $\cup_{j>j_0}\Gamma_1(z_j)$. The first is contained in a ball of radius h, since $h - y_{j_0} < h$. The second is contained in $B(0, (\beta+1) \cdot h)$. q.e.d.

The construction of a sequence $\Sigma = \{z_n\}_{n \in \mathbb{N}} \subset \mathcal{C}$ with the properties stated above is achieved by the method illustrated in Figs. 2.3 and 2.4. Start from a point $z \in \mathcal{C}$, $z \equiv (x_1, y_1) = (x_1, \phi(x_1))$ and consider the intersection ζ_1 of the horizontal line through z, in the (x,y)-plane, with the boundary of the cone $\Gamma_\beta(0)$; then we move along the vertical line through ζ_1 until we meet the curve $\mathcal{C}$ at the point (x_2, y_2), and so on. A similar construction applies for curves converging to a point $x \neq 0$: It suffices first to translate to the origin, and then to translate back to x. (Complete details can be found in [Sue86, p. 663]).

This concludes the proof of the proposition. q.e.d.

Denote by $\Sigma(z, \beta, \mathcal{C})$ the sequence constructed in Proposition 2.23. In view of Claim 2.24, it follows that the constant for the cross-section condition for the region Σ_{Γ_1} is independent of the point $z \in \mathcal{C}$ and also independent of the curve $\mathcal{C}$. The following example is based on this observation.

Example 2.25 *There is an approach family $\mathcal{L}$ in the upper half-space with the following properties:*

1. *$\mathcal{L}$ is not invariant under translations;*

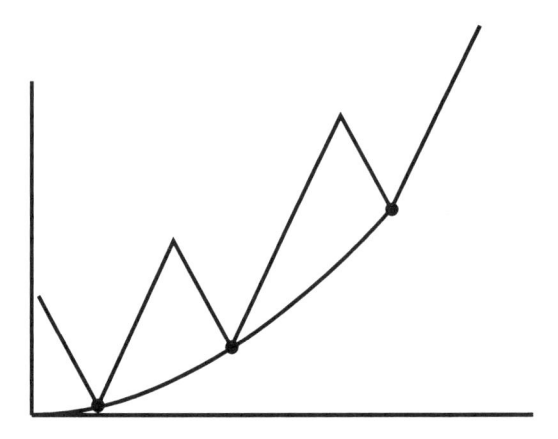

Figure 2.4: The Nagel–Stein approach region is the union of the cones at the points (x_n, y_n).

2. the cross-section condition for $\mathcal{L}$ is satisfied, with a constant uniform for all points;

3. the tent condition for $\mathcal{L}$ is not satisfied.

It suffices to move the point z upon which the construction of $\Sigma(z, \beta, \mathcal{C})$ depends. Thus, we fix a curve $\mathcal{C}$, say the one given by $y = x^2$. Fix $\beta = 1$. Let $\mathcal{G} = \Gamma_{1/10}$. Consider the Carleson tent $\triangle^{\mathcal{G}}(B)$ where B is chosen in such a way that the vertex of $\triangle^{\mathcal{G}}(B)$ is the point $(1,1)$ (the vertex of $\triangle^{\mathcal{G}}(B)$ is the unique point ς such that $\mathcal{G}^{\downarrow}(\varsigma) = B$). The upper boundary of $\triangle^{\mathcal{G}}(B)$ consists of two line segments. Let ℓ be the one with positive slope. For $0 < x < 1 - 1/10$, let p_x be the point at which the curve $x + \mathcal{C}$ intersects the segment ℓ. Let $\mathcal{L}(x)$ be the completion, with respect to Γ_1, of the sequence

$$\Sigma(p_x, \beta, x + \mathcal{C}).$$

Then $\mathcal{L}$ has the required properties.

The point of the previous example is the following: Without translation invariance, a pointwise bound on the measure of the cross-sections of the approach regions is *not* sufficient to yield the subordination with respect to the cones.

Consider the unit disc again. We claim that Theorem 2.14 (the equivalence between the tent condition and the distributional inequality for approach families) and Theorem 1.19 (the equivalence between almost everywhere pointwise convergence for a sequence of operators and weak type estimates for the associated maximal operator) provide a clear way to see that Poisson integrals of functions in $L^1(bD)$ do not converge

2.2. THE NAGEL–STEIN APPROACH REGIONS

along any along any rotation invariant approach family that is generated by a tangential curve in the unit disc. First, recall that a continuous curve $y = f(x), x \in [0, 2\pi]$ is tangential at 0 if $\lim_{x \to 0} \frac{f(x)}{x} = 0$. Assume for simplicity that f is increasing, and that $0 < f(x) < 1$. Let $L_1 \equiv \{(1 - f(x))e^{ix} : x \in (0, 2\pi)\}$. Let L_0 be the completion of L_1 with respect to the family of cones. It is then clear that the cross-section condition for L_0 fails. Let $\mathcal{L}$ be the approach family generated by L_0 by rotations in the unit disc. Since in this case, by Theorem 2.22, the tent condition specializes to the cross-section condition, Theorem 1.19 implies that there is a function in $L^1(\partial D)$ whose Poisson extension *diverges* along the given tangential curve for almost every point in bD. While this result is not as strong as Theorem 1.12 or as the version given by A. Zygmund and others ([Aik91], [LP]), this technique has the advantage of being conceptually clearer than the original proof given by Littlewood. Cf. [Sjö96]. See also [Rud79], [Rud88].

Aside: Overconvergence of Power Series

There is a resemblance between the Nagel–Stein phenomenon and the *overconvergence* of power series. The sequence of partial sums of a power series $\sum_0^\infty a_j z^j$ does not converge at any point z that lies outside its circle of convergence, but a *subsequence* of the sequence of partial sums may very well converge (uniformly) at such points. This phenomenon, called **overconvergence** by A. Ostrowski, is linked to the **lacunarity structure** of the coefficients of the power series. More precisely, consider power series with an infinite number of **lacunary gaps**, i.e., power series of the form

$$\sum_j a_j z^j, \qquad a_j = 0 \qquad \text{for all} \qquad m_k < j < m'_k,$$

where the two sequences of integers m_k and m'_k have the property that

$$\frac{m'_k}{m_k} > 1 + \varepsilon \qquad (2.14)$$

for a fixed $\varepsilon > 0$. Assume that the radius of convergence is 1. J. Hadamard [Had92] examined the special case in which $m'_k \equiv m_{k+1}$, so that each block of monomials actually occurring in the series consists of a single monomial $a_{m_k} z^{m_k}$ (as in the series $\sum_j z^{2^j}$) and proved that every point z in the circle of convergence $|z| = 1$ is singular. A. Ostrowski [Ost26, p. 254] observed the paradox of this theorem: "in it just the power series with gaps, which are particularly rapidly convergent, are asserted to be non-continuable, that is to say to possess particularly heavy singularities on the circle of convergence", and interpreted Hadamard's theorem in terms of a more general one, which he illustrated with the example of the series

$$\sum_{j=1}^{\infty} \frac{[z(1-z)]^{4^j}}{b_j}, \qquad (2.15)$$

where b_j is a coefficient of maximal modulus in the polynomial $[z(1-z)]^{4^j}$ [Ost26, pp. 252–253]. The monomials in each term $[z(1-z)]^{4^j}$ have degrees varying between 4^n and $2 \cdot 4^n$; therefore, monomials in different terms do not overlap. Let

$$\sum_{j} c_j z^j \qquad (2.16)$$

be the power series obtained by expanding the terms

$$\frac{[z(1-z)]^{4^j}}{b_j}$$

and writing them down in their natural order. Then the radius of convergence of (2.16) is 1, since each coefficient in (2.16) is bounded by 1, and infinitely many of them are equal to 1; in particular, the series (2.16) does not converge at the point 1. Observe that (2.15) is the subsequence of the sequence of partial sums of (2.16) obtained as $\sum_{j=4}^{2 \cdot 4^n} c_j z^j$. Moreover, the series (2.15) is invariant under the transformation $z \mapsto 1 - z$. In particular, it is uniformly convergent in a neighborhood of 1. Thus, the point 1 is regular for the series (2.16), and the subsequence $\sum_{j=4}^{2 \cdot 4^n} c_j z^j$ obtained by taking the last term of each lacunary block is uniformly convergent in a neighborhood of 1; in particular, it provides an analytic continuation for the function. A. Ostrowski (1921, see references in [Ost26]) showed that this is a general fact and proved the following result (which implies Hadamard's theorem): If a power series $f = \sum_m a_m z^m$ has an infinite number of lacunary gaps between $\{m_k\}$ and $\{m'_{k+1}\}$, then the subsequence of partial sums

$$s_{m_k} = \sum_{j=0}^{m_k} a_j z^j$$

is uniformly convergent in an open neighborhood of a *regular* point of the function f in the circle of convergence (if there is any regular point). He also observed that "a new light is perhaps thrown on [the paradox

2.3. GOALS, PROBLEMS AND RESULTS

described before] by our theorem. We could say that it is just because of the particularly rapid convergence of Hadamard's series inside the circle of convergence that it would converge also outside it, were this circle not the natural boundary for the function." [Ost26, p. 254]. Later A. Zygmund gave a new proof of Ostrowski's theorem [Zyg31]. See also [Bou37].

2.3 Goals, Problems and Results

If $\{\mathcal{G}_\alpha\}_{\alpha \in I}$ is an approach system for (D, W), then our goal is the construction of an approach family $\mathcal{L}$ for (D, W), whose support $\mathrm{supp}(\mathcal{L})$ is a set of full measure in W and such that

(**tent condition**) $\mathcal{L}$ is subordinate to one particular $\mathcal{G}_{\bar{\alpha}}$ (and therefore any $\mathcal{G}_\alpha$); we saw in Theorem 2.9 that then there is convergence along $\mathcal{L}$ for the same class of functions that converge along $\mathcal{G}_{\bar{\alpha}}$;

(**exotic**) $\mathcal{L}$ is exotic with respect to $\{\mathcal{G}_\alpha\}_\alpha$ at each point of a subset of full measure of W.

Whenever these conditions hold, the approach family $\mathcal{L}$ is a **Nagel–Stein approach family** relative to $\{\mathcal{G}_\alpha\}_{\alpha \in I}$.

In the setting of the Euclidean half-spaces $\mathbb{R}^{n+1}_+$, for a translation invariant approach family $\mathcal{L}(w) = w + \mathbf{L}$, the tent condition for $\mathcal{L}$ (which is always equivalent to the subordination of $\mathcal{L}$ by $\mathcal{G}_\alpha$) is equivalent to the cross-section condition for $\mathbf{L} = \mathcal{L}(0)$ (but in general, without group invariance, this equivalence does not hold). J. Sueiro has shown that this situation also holds in the case where the space of approach D to W is given by the product of W with $(0, \infty)$:

$$D \stackrel{\mathrm{def}}{=} W \times (0, \infty)$$

under the additional assumptions that W is a group, the measure ν is invariant, and the metric is invariant; in particular,

$$|\mathbb{B}(w, r)| = |\mathbb{B}(v, r)| \qquad (2.17)$$

for all $w, v \in W$ and $r > 0$. In this setting, the tent condition is equivalent to a cross-section condition, as in the Euclidean half-space, and the construction of an exotic approach family proceeds along the lines of Section 2.2. This method can be applied to the unit ball in $\mathbb{C}^n$ (where the natural approach regions for bounded holomorphic functions are the *admissible regions*; see Chapter 3), since then there is indeed a group acting on the boundary in the prescribed way: the Heisenberg group. Then the construction of an exotic approach family that also satisfies the tent condition is achieved by first constructing at one point, and then applying the motions of the group in order to define it at other points, as in $\mathbb{R}^{n+1}_+$. See [Sue86] for more details.

A further extension has been given by M. Andersson and H. Carlsson to the setting of a space of homogeneous type that is a differentiable manifold, such as the boundary of smoothly bounded domains in $\mathbb{R}^n$ (where the natural approach regions for bounded harmonic functions are cones) and the boundary of strongly pseudoconvex domains in $\mathbb{C}^n$ (where the natural approach regions for bounded holomorphic functions are the *admissible* regions introduced in [Ste72]). The result of M. Andersson and H. Carlsson is based on the existence of a *pseudogroup* of local diffeomorphisms preserving the family of balls, *modulo equivalence*, under the assumption that the approach regions are connected to each other by the smooth family of diffeomorphisms. In this setting, a cross-section condition will also control the distributional inequality. In particular, this hypothesis implies that

$$|\mathbb{B}(w,r)| \simeq |\mathbb{B}(v,r)|, \, w, v \in W, r > 0. \tag{2.18}$$

See [AC92] for more details. In particular, in these settings, one may also first construct an approach region that satisfies the cross-section condition at one fixed point, and then move it to nearby points, using the motions of the pseudogroup of local diffeomorphisms.

There are important, indeed typical, cases, in which this method of constructing exotic approach regions of convergence is faced with a fundamental difficulty, namely the absence of a group (or pseudo-group) of (local) transformations; in particular, (2.17) and (2.18) fail for:

1. NTA domains in $\mathbb{R}^n$, such as the von Koch snowflake, with respect to the harmonic measure; there the natural approach regions for the almost everywhere convergence of bounded harmonic functions have been described by D. Jerison and C. Kenig as *twisted cones* (see the next chapter); in general, the harmonic measure does not satisfy (2.18);

2. pseudoconvex domains of finite type in $\mathbb{C}^n$, such as the egg domain

$$\{(z,w) \in \mathbb{C}^2 : |z|^2 + |w|^4 < 1\},$$

for which the natural approach regions for bounded holomorphic functions, and the relevant metric in the boundary, have been identified by A. Nagel, E.M. Stein and S. Wainger [NSW81], [NSW85].

2.3. GOALS, PROBLEMS AND RESULTS

3. a non-homogeneous tree, endowed with the potential theory generated by a nearest-neighborhood random walk (under certain regularity assumptions); see the next chapter.

In particular, in all these cases (2.17) and (2.18) fail, and there is no group or pseudogroup of local diffeomorphisms at our disposal.

It is of independent, intrinsic interest to see whether the existence of approach regions of Nagel–Stein type is independent of the existence of a large group of automorphisms of the given structure. In Chapters 5 and 6 we show that this question has an affirmative answer, and that the Nagel–Stein phenomenon is linked to the fundamental structure of space of homogeneous type, rather than to a group structure.

We present a new technique to construct exotic approach regions of convergence, which does not depend on the existence of a "translation" structure. The main idea is first to solve the discrete case of a tree (Chapter 4), and then to apply to the given space of homogeneous type the *quasidyadic decomposition in cubes* found by M. Christ for spaces of homogeneous type (cf. [Chr90]). To each such decomposition there is associated a tree, which encodes the inclusion relations between the *cubes* of the decomposition; this tree can be embedded in the domain, and this embedding enables us to transfer to the domain the approach regions constructed for a tree; the final, crucial step consists of showing that the resulting approach regions are indeed exotic and satisfy the tent condition *with respect to the domain*.

Symmetric spaces and products of Euclidean half-spaces offer other kinds of difficulties; see [Sve95], [Sve96a], [Sve96b].

Chapter 3

The Geometric Contexts

3.1 NTA Domains in $\mathbb{R}^n$

A domain for which the Dirichlet problem has a solution for every continuous boundary datum is called a **regular** domain for the Dirichlet problem.

Domains that are not regular for the Dirichlet problem were discovered in the first decade of the 20th century. Zaremba showed that there is no harmonic function on the punctured disc $D\setminus\{0\}$ with boundary values 1 at the origin and 0 at the boundary bD; Lebesgue gave an example of a nonregular domain that is homeomorphic to the ball, and smooth except at one point.

N. Wiener looked at the example of the punctured disc $D\setminus\{0\}$, and showed that one could profitably analyze the problem in two separate components [Wie24]. First, one should still try to associate a harmonic function to each continuous boundary data, in such a way that the **generalized solution** that one seeks specializes to the classical solution, when the domain is regular. Secondly, one is asked to study the boundary behaviour of the generalized solution, and find conditions that guarantee that the generalized solution does converge to the pre-assigned data. In fact, if we consider the smaller domain $\varepsilon < |z| < 1$ inside the punctured disc, and extend continuously the given boundary condition, then we do obtain the solution

$$u(r,\theta) = \frac{\log r}{\log \varepsilon}$$

for the Dirichlet problem; for every point in the punctured disc, the limit of $u(r,\theta)$ for $\varepsilon \to 0$ is the function identically zero, which does not assume the required limit at the boundary. N. Wiener interpreted this example in a positive way, as showing "not [the] non-existence of a harmonic function corresponding to the boundary conditions, but [only that] this harmonic function fails to assume continuously the desired value at the origin" [Wie24, p. 25]. He showed that, for any bounded domain D in $\mathbb{R}^n$, there is indeed a harmonic function H_f corresponding to a given continuous boundary function f. The method used by N. Wiener is the following. First, we extend the function f to a function ϕ defined and continuous on the closure of the domain. Then we construct an increasing sequence of regular domains $\{D_n\}_n$ contained in the domain D, whose union is equal to the whole domain. Next, we show that the solution corresponding to the restriction of ϕ on the boundary of D_n tends locally uniformly in D to a harmonic function in D, as $n \to \infty$. Finally, we show that this limit function does not depend on the choices made during this construction. The generalized solution H_f lies between the upper and lower bounds of the given function f and is at any point of D an additive functional of the boundary values, i.e. for any fixed $z_0 \in D$, the function $P_{z_0} : C(\mathrm{b}D) \to \mathbb{R}$, which maps $f \in C(\mathrm{b}D)$ into $P_{z_0}(f) \stackrel{\text{def}}{=} H_f(z_0)$, is a positive linear functional. By the Riesz representation theorem, there is a measure ν^{z_0} on $\mathrm{b}D$ such that

$$P_{z_0}(f) = H_f(z_0) = \int_{\mathrm{b}D} f(w)\, d\nu^{z_0}(w)\,.$$

The measure ν^{z_0} is the **harmonic measure** of the domain D. Cf. [Car82].

The notion of nontangentially accessible domain was introduced by D.S. Jerison and C.E. Kenig in 1982. Before giving the precise definition, let us recall two basic geometric constructions that are linked to each other. As we saw in Lemma 1.15, any open set in a space of homogeneous type admits a Whitney-type decomposition into a collection of balls each of whose radius is comparable to its distance from the boundary of the open set; moreover, the number of balls (in the collection) that intersect is bounded above by a universal constant. In $\mathbb{R}^n$, this decomposition takes the particularly simple form of a decomposition in disjoint open dyadic cubes, which cover all of the given open set with the exception of a set of measure zero and whose radius is comparable to their distance from the boundary of the open set [Ste70, pp. 167–168]. We have used the Whitney-type decomposition to reduce a distribution function inequality to a tent condition. On the other hand, for each w in $\mathbb{R}^n$, thought of as the boundary of $\mathbb{R}^{n+1}_+$, and each $r > 0$, there is one point $z(w,r) \in \mathbb{R}^{n+1}_+$ whose distance from $\mathbb{R}^n$ is comparable to its distance from w and to r: The point $z(w,r)$ is simply the one belonging to the normal segment issuing from w at distance r from $\mathbb{R}^n$, i.e. $z(w,r) = (w,r) \in \mathbb{R}^{n+1}_+$. This simple geometric fact was also exploited by A. Zygmund, together with

3.1. NTA DOMAINS IN $\mathbb{R}^N$

the dyadic decomposition of the interval $[0, 1]$, in his real variable proof of Littlewood's Theorem (Theorem 1.12; see [Zyg49]). The geometric fact that is involved here is possibility of associating a ball of Whitney-type in the domain to each ball (of small radius) on the boundary.
Let
$$B(z, r) = \{w \in \mathbb{R}^n : |z - w| < r\}$$
for $z \in \mathbb{R}^n$ and $r > 0$. Let $d(S_1, S_2)$ be the Euclidean distance between two subsets S_1 and S_2 of $\mathbb{R}^n$, i.e.
$$d(A, B) = \inf\{|z - w| : z \in S_1, w \in S_2\}.$$
Let m be a positive real number. An m-**Whitney ball** for a domain D in $\mathbb{R}^n$ is a ball $B(z, r)$ contained in D such that its radius is comparable, via m, to its distance from the boundary bD, i.e.
$$\frac{1}{\mathtt{m}} \cdot r < d(B(z, r), \mathrm{b}D) < \mathtt{m} \cdot r. \tag{3.1}$$

If z_1 and z_2 are points in $D \subset \mathbb{R}^n$, then an m-**Harnack chain** from z_1 to z_2 is a sequence of m-Whitney balls such that the first ball contains z_1, the last ball contains z_2, and consecutive balls intersect.

A bounded domain D in $\mathbb{R}^n$ with boundary bD is called **nontangentially accessible**, or **NTA**, if there are constants m and r_0 such that

1. D **satisfies the corkscrew condition**: For each point $w \in \mathrm{b}D$ and each positive $r < r_0$, there exists a point $z \equiv z(w, r)$ in D such that
$$\frac{1}{\mathtt{m}} \cdot r < |z - w| < r \tag{3.2}$$
and
$$d(z, \mathrm{b}D) > \frac{1}{\mathtt{m}} \cdot r; \tag{3.3}$$

2. $\mathbb{R}^n \setminus D$ satisfies the corkscrew condition;

3. D **satisfies the m-Harnack chain condition**: If $\varepsilon > 0$ and z_1 and z_2 belong to D, $d(z_j, \mathrm{b}D) > \varepsilon$ and $|z_1 - z_2| < C \cdot \varepsilon$, then there exists an m-Harnack chain from z_1 to z_2 whose length depends on C but not on ε.

Observe that, by Harnack's inequality, if u is a positive harmonic function on D and there is an m-Harnack chain from z_1 to z_2, then $\frac{1}{c} \cdot u(z_2) < u(z_1) < c \cdot u(z_2)$, where c depends only on m and on the length of the Harnack chain between z_1 and z_2.

D.S. Jerison and C.E. Kenig showed that these geometric constraints are sufficient to obtain a large class of domains to which the results on the boundary behaviour of harmonic functions on $\mathbb{R}^{n+1}_+$ can be extended. In particular, they showed that the doubling condition holds for the harmonic measure of an NTA domain. Let $\nu \equiv \nu^{z_0}$ for some fixed point $z_0 \in D$.

Theorem 3.1 *Let D be an NTA domain. Then $\mathrm{b}D$ is a space of homogeneous type with respect to the harmonic measure and the induced Euclidean metric. NTA domains are regular for the Dirichlet problem.*

This is [JK82, p. 93; Lemma 4.1, p. 97; Lemma 4.9, p. 101].

Let D be an NTA domain. The **nontangential approach region** (or **corkscrew**) for D at $w \in \mathrm{b}D$ is the set

$$\Gamma_\alpha(w) = \{z \in D : |z - w| < (1+\alpha)d(z, \mathrm{b}D)\} \qquad (3.4)$$

for $\alpha > 0$. In order to get an idea of the shape of these approach regions Γ_α, let us examine the case of the von Koch snowflake D, bounded by the closed von Koch curve $\mathrm{b}D$ [vK06]. The von Koch snowflake domain D is defined as the union of an increasing sequence $\{D_n\}_n$ of domains bounded by polygons. The first domain D_1 is an equilateral triangle. The domain D_{n+1} is obtained from D_n by completing the middle third of each side of D_n to an open equilateral triangle pointing outside D_n (see [Pom92, p. 111], [Sag94, p. 145–149] or [LV73, p. 104]). The von Koch snowflake is an NTA domain. In fact, it is a *quasicircle*; cf. [LV73, p. 104] and [JK82, Theorem 2.7, p. 91]. The boundary of the von Koch snowflake is nonrectifiable; NTA domains need not have a Lipschitz boundary. We are interested in the behaviour of the approach region Γ_α at a point that belongs to the support of the harmonic measure for D. The harmonic measure for the von Koch snowflake is supported in the set of **twist points**, i.e. those points of the boundary that can only be the limit of a path in the domain that "winds around the limiting point infinitely often in both directions" [Pom92, p. 127] and [DBFU]; cf. [Bur89]. In particular, the corkscrew at a twist point of the snowflake domain has a twisted shape.

Proposition 3.2 *For $\alpha > 0$, the nontangential approach regions Γ_α given in (3.4) define an approach system for $(D, \mathrm{b}D)$.*

Proof. For $w \in \mathrm{b}D$ and $r > 0$ let

$$\mathbb{B}(w, r) \stackrel{\text{def}}{=} B(w, r) \cap \mathrm{b}D.$$

3.2. DOMAINS IN $\mathbb{C}^N$

For $z \in D$ let
$$d(z) \stackrel{\text{def}}{=} \min_{w \in bD} d(w, z)$$
and let $\pi(z) \in bD$ be a point of minimal distance from z, so that $d(\pi(z), z) = d(z)$. Then
$$\frac{1}{2} \cdot \mathbb{B}(\pi(z), d(z)) \subset \Gamma_1^\downarrow(z) \subset 3 \cdot \mathbb{B}(\pi(z), d(z))$$
and
$$\frac{\alpha}{2} \cdot \mathbb{B}(\pi(z), d(z)) \subset \Gamma_\alpha^\downarrow(z) \subset (2 + \alpha) \cdot \mathbb{B}(\pi(z), d(z)) .$$

q.e.d.

For any two points $z, z_0 \in D$, the harmonic measure ν^z is absolutely continuous with respect to ν^{z_0}. The **kernel function** K is defined to be the Radon–Nikodym derivative
$$K(z, w) = \frac{d\nu^z}{d\nu^{z_0}}(w)$$
where $z \in D$, $w \in bD$. Theorem 1.1 implies that the kernel function for the unit disc is the Poisson kernel.

Theorem 3.3 *Let D be an NTA domain. For $f \in L^1(W, d\nu)$ and $z \in D$, let*
$$Kf(z) \stackrel{\text{def}}{=} \int_{bD} f(w) K(z, w) d\nu(w) .$$
Then Kf converges to f along Γ_α on a subset of full harmonic measure in bD, for each $\alpha > 0$. Each bounded harmonic function in D converges along Γ_α on a set of full harmonic measure in bD.

This is [JK82, Theorem 5.8, p. 105; Theorem 6.4, p. 112].

3.2 Domains in $\mathbb{C}^n$

A complex-valued function defined in an open subset of $\mathbb{C}^n$ is **holomorphic** if it is satisfies the Cauchy–Riemann equations in each variable separately. The problem of constructing holomorphic functions that have singularities at the boundary of a domain in $\mathbb{C}^n$ (the *Levi problem*) is more complex for $n > 1$, since then $\mathbb{C}^n$ contains domains with the property that all holomorphic functions defined in the domain admit a holomorphic extension to a larger domain, [Har06a], [Har06b] (see [Kra92a]). It is therefore natural to restrict ourselves to a **domain of holomorphy**, i.e. a domain that admits a function, holomorphic in the domain and *singular* at every boundary point.

The Unit Ball in $\mathbb{C}^n$

The unit ball in $\mathbb{C}^n$ is the simplest domain of holomorphy [Lev11] different from the polydisc D^n [Poi07].

A useful kernel function for the study of the boundary behaviour of holomorphic functions in the unit ball

$$D \stackrel{\text{def}}{=} \{z \in \mathbb{C}^n : |z| < 1\}$$

is the Poisson–Szegö kernel $\mathcal{P}$,

$$\mathcal{P}(z, w) \stackrel{\text{def}}{=} c_n \frac{(1 - |z|^2)^n}{|1 - z \cdot \overline{w}|^{2n}}, \quad z \in D, \ w \in \mathrm{b}D$$

where

$$z \cdot \overline{w} \stackrel{\text{def}}{=} \sum_{j=1}^{n} z_j \cdot \overline{w_j},$$

$\overline{\zeta}$ is the conjugate of $\zeta \in \mathbb{C}$, $|\zeta|$ is the norm of $\zeta \in \mathbb{C}$ and $c_n = \frac{(n-1)!}{2\pi^n}$. The Poisson–Szegö kernel is the kernel function for the functions in D that are harmonic with respect to the Laplace–Beltrami operator for the Bergmann metric of D [Rud80]. Every continuous function f on $\overline{D}$ that is holomorphic in D is the Poisson–Szegö integral of its boundary values:

$$f(z) = \mathcal{P}[f](z) \stackrel{\text{def}}{=} \int_{\mathrm{b}D} \mathcal{P}(z, w) f(w) d\nu(w),$$

where ν is the $(2n - 1)$-dimensional Hausdorff measure of $\mathrm{b}D$ [Kra92a, Proposition 1.5.1].

The boundary $\mathrm{b}D$ has another notable metric ρ, in addition to that induced by the Euclidean metric d of the ambient space. A proof of the following lemma can be found in [Kra92b] and [Rud80, p. 66].

Lemma 3.4 *The expression*

$$\rho(w, u) \stackrel{\text{def}}{=} |1 - w \cdot \overline{u}|^{1/2}$$

defines a metric on $\mathrm{b}D$.

The expression

$$|1 - w \cdot \overline{u}|$$

has the following geometric meaning. First, we may identify the tangent space $T_u(\mathrm{b}D)$ to $\mathrm{b}D$ at a point $u \in \mathrm{b}D$ (a real vector space) with a subspace of $\mathbb{C}^n$ (a complex vector space). Observe that the maximal *complex*

3.2. DOMAINS IN $\mathbb{C}^N$

subspace $T_u^c(bD)$ of the tangent space to bD at u has codimension one in $T_u(bD)$. In fact, the "missing direction" is generated (over the reals) by the vector $i \cdot N_u$, where N_u is the unit normal vector to the hypersurface bD at the point u, and $i\cdot$ is multiplication by i in $\mathbb{C}^n$. On the sphere bD,

$$N_u \equiv u.$$

See Figs. 3.1 and 3.2. Therefore, the tangent space $T_u(bD)$ splits as an orthogonal direct sum, over the reals, of $T_u^c(bD)$ with $\mathbb{R} \cdot (i \cdot N_u)$. The complex subspace $T_u^c(bD)$ of $\mathbb{C}^n$ is the **complex tangent space**. In particular, the Euclidean distance

$$d(z, u + T_u^c(bD))$$

from a point z to the affine subspace $u + T_u^c(bD)$, obtained by translating to u the complex tangent space at u, is given by

$$d(z, u + T_u^c(bD)) = |\langle z - u, u \rangle u|,$$

where $\langle \cdot, \cdot \rangle$ denotes the inner product in $\mathbb{C}^n$, i.e. $\langle z, u \rangle \equiv z \cdot \overline{u} \equiv \sum_{j=1}^n z_j \overline{u_j}$. Since $|u| = 1$, we obtain

$$d(z, u + T_u^c(bD)) = |\langle z - u, u \rangle u| = |\langle z - u, u \rangle| = |1 - z \cdot \overline{u}|.$$

This interpretation of the metric ρ has the notable implication that the natural approach regions, for bounded holomorphic functions in the unit ball, have a parabolic order of contact with the boundary bD in the complex tangent directions. Consider the approach family $\mathcal{A}_\alpha(w)$ of the **admissible** approach regions, defined as

$$\mathcal{A}_\alpha(u) \stackrel{\text{def}}{=} \{z \in D : |1 - \langle z, u \rangle| < \alpha \cdot d(z)\},$$

where $d(z) = 1 - |z|$ is the Euclidean distance from z to the boundary. The geometric interpretation of $|1 - \langle z, u \rangle|$ implies that the approach family $\mathcal{A}_\alpha$ is nontangential in the missing direction, but parabolic tangential in the complex tangential directions. See [Rud80, p. 73] for the explicit computation. (See Figs. 3.3 and 3.4.) When the point z lies in the missing direction (with respect to the point u, i.e., the vector $z - u$ is a real multiple of $i \cdot N_u$) then the intersection of the approach region $\mathcal{A}_\alpha(u)$ with the plane $u + \mathbb{C}N_u$ through u generated by N_u and $i \cdot N_u$ is given by

$$\sec_{N_u} = \{z \in D \cap (u + \mathbb{C}N_u) : |z - u| \leq \alpha d(z)\}$$

which is nontangential. When the point z lies in the complex tangential direction with respect to u (i.e., the vector $z - u$ belongs to the complex tangent space $T_u^c(bD)$), then the intersection of the approach region

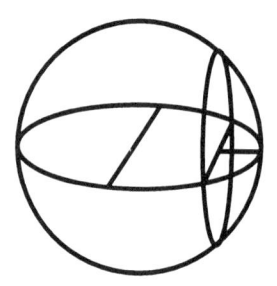

Figure 3.1: The unit ball.

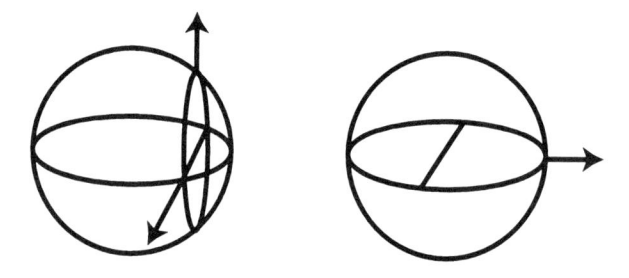

Figure 3.2: The complex tangential directions (left); the normal direction (right).

$\mathcal{A}_\alpha(u)$ with any plane $u+\mathbb{C}\tau$ through u generated by a complex tangential vector τ is given by

$$\sec_\tau = \{ z \in D \cap \tau : d(z, u + \mathbb{C}\tau) \leq \alpha d(z) \}$$

which is tangential to the boundary bD. Moreover, the order of contact of $\sec_\tau$ depends on the order of contact of the boundary bD with the complex tangent plane in the direction of τ. For the ball, this order of contact is quadratic (see Fig. 3.5), but for the point $u = (1,0) \in \mathbb{C}^2$ in the domain D in $\mathbb{C}^2$ given by

$$D = \{ (z, \varsigma) \in \mathbb{C}^2 : |z|^2 + |\varsigma|^4 < 1 \} ,$$

the boundary bD has a fourth order of contact with the complex direction $(0,1)$ at u. We will return on this topic in the next section.

A proof of the following theorem can be found in [Rud80, Chapter 5, p. 75].

3.2. DOMAINS IN $\mathbb{C}^N$

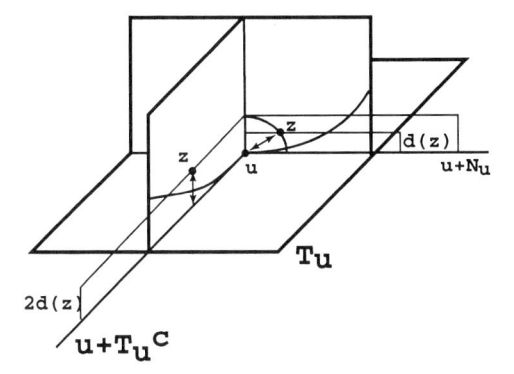

Figure 3.3: The admissible approach region for the unit ball. Arrows represent $d(z, u + T_u^c(\mathrm{b}D))$.

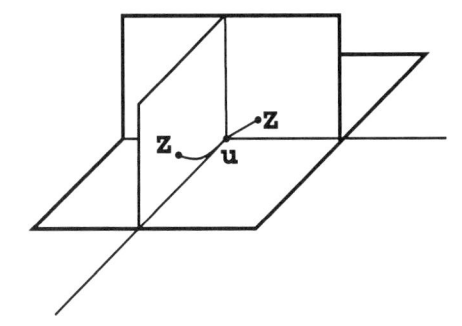

Figure 3.4: The admissible approach region for the unit ball: nontangential in the missing direction, parabolic in the others.

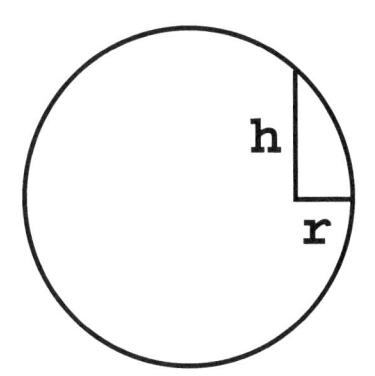

Figure 3.5: The relative size of the largest embedded analytic disc in the complex tangential direction for the unit ball: $h = r^{1/2}(2-r)^{1/2}$.

Theorem 3.5 *Let D be the unit ball in $\mathbb{C}^n$, ν the Hausdorff $(2n-1)$-dimensional measure on $\mathrm{b}D$, ρ the anisotropic metric defined in Lemma 3.4. Then*

1. *the doubling property holds for ν with respect to the metric ρ, thus $(\mathrm{b}D, \nu, \rho)$ is a space of homogeneous type;*

2. *the collection $\{\mathcal{A}_\alpha\}_{\alpha>1}$ forms an approach system;*

3. *the following pointwise estimate holds for any each $f \in L^1(W)$ and $w \in W$*
$$\sup_{z \in \mathcal{A}_\alpha(w)} \mathcal{P}[f](z) \leq c_\alpha M' f(w); \qquad (3.5)$$

4. *each bounded holomorphic function on the unit ball in $\mathbb{C}^n$ converges along $\mathcal{A}_\alpha$ on a subset of full measure of the boundary.*

Part 4 of Theorem 3.5 also holds for holomorphic functions in the Hardy classes and in the Nevanlinna class; cf. [Rud80, Theorem 5.6.4, p. 85] and [Kor69]. A theorem of Littlewood type has been proved in this context by M. Hakim and N. Sibony; for certain approach regions which are *larger* than $\mathcal{A}_\alpha$, they showed the existence of bounded holomorphic functions that fail to converge along the larger regions, for almost every point; see [HS83].

Complex(ified) Tangent Spaces

Since we are dealing with complex-valued functions f and first-order differential operators like

$$\frac{\partial}{\partial z_k} f \stackrel{\text{def}}{=} \frac{1}{2}\left(\frac{\partial}{\partial x_k} - i\frac{\partial}{\partial y_k}\right) f \qquad (3.6)$$

and

$$\frac{\partial}{\partial \bar{z}_k} f \stackrel{\text{def}}{=} \frac{1}{2}\left(\frac{\partial}{\partial x_k} + i\frac{\partial}{\partial y_k}\right) f,$$

in order to avoid a confusion between the multiplication by i in $\mathbb{C}^n$ and the multiplication by i in (3.6), we denote the former by $N \in \mathbb{C}^n \mapsto J(N) \in \mathbb{C}^n$, so that the missing direction is generated by $J(N_u)$. Observe that the first-order differential operator $\frac{\partial}{\partial z}$ is not an element of the tangent space $T(\mathbb{C}^n)$ to $\mathbb{C}^n$, but an element of its complexification $\mathbb{C} \otimes T(\mathbb{C}^n)$: Formally, one has $i\frac{\partial}{\partial y} \equiv i \otimes \frac{\partial}{\partial y}$. The operation of forming the tensor product with

3.2. DOMAINS IN $\mathbb{C}^N$

$\mathbb{C}$ means that we admit complex coeffients rather than only real ones. In particular, $\mathbb{C} \otimes T(\mathbb{C}^n)$ is the direct sum (over $\mathbb{R}$)

$$\mathbb{C} \otimes T(\mathbb{C}^n) = T(\mathbb{C}^n) \oplus i \cdot T(\mathbb{C}^n),$$

so that vectors in $\mathbb{C} \otimes T(\mathbb{C}^n)$ have a "real part" and an "imaginary part". Moreover, the conjugate

$$\overline{a + i \otimes b}$$

is defined as $a - i \otimes b$. On the other hand, the operator J extends in a natural way to the complexified tangent space $\mathbb{C} \otimes T(\mathbb{C}^n)$, by letting

$$J(\zeta \otimes v) \stackrel{\text{def}}{=} \zeta \otimes J(v), \text{ for } \zeta \in \mathbb{C}, v \in T(\mathbb{C}^n),$$

and this extension defines a *second* structure of $\mathbb{C}$-vector space for $\mathbb{C} \otimes T(\mathbb{C}^n)$, beside the intrinsic one given by $i \cdot (a + i \otimes b) \stackrel{\text{def}}{=} 1 \otimes (-b) + i \otimes a$, for $a, b \in T(\mathbb{C}^n)$. We have therefore *two* structures of $\mathbb{C}$-vector space for $\mathbb{C} \otimes T(\mathbb{C}^n)$, which *agree* on a subspace $T^{(1,0)}(\mathbb{C}^n) \subset \mathbb{C} \otimes T(\mathbb{C}^n)$, namely on

$$T^{(1,0)}(\mathbb{C}^n) \stackrel{\text{def}}{=} \{u \in \mathbb{C} \otimes T(\mathbb{C}^n) : J(u) = i \cdot u\}.$$

Vectors of the form

$$v - i \otimes J(v) \equiv 1 \otimes v - i \otimes J(v) \equiv v - iJ(v)$$

belong to $T^{(1,0)}(\mathbb{C}^n)$. The converse is also true. In particular, the assignment

$$v \in T(\mathbb{C}^n) \mapsto V = \frac{1}{2}(v - i \otimes J(v)) \in T^{(1,0)}(\mathbb{C}^n)$$

is an isomorphism of $\mathbb{C}$-vector spaces. For example, in $\mathbb{C}$, corresponding to $\frac{\partial}{\partial x}$ there is associated the vector $\frac{1}{2}(\frac{\partial}{\partial x} - i \otimes J(\frac{\partial}{\partial x}))$. Since $J(\frac{\partial}{\partial x}) \equiv J(1) = i \cdot 1 \equiv i \equiv (0,1) \equiv \frac{\partial}{\partial y}$, we obtain, corresponding to $\frac{\partial}{\partial x}$, the operator $\frac{\partial}{\partial z}$ of (3.6). In fact, for holomorphic functions f, one has $\frac{\partial}{\partial x} f = \frac{\partial}{\partial z} f$, and the complex vector space $T(\mathbb{C}^n)$, which we have seen to be isomorphic to the complex vector space $T^{(1,0)}(\mathbb{C}^n)$, can also be described as the vector space of all complex derivations of the algebra of germs of holomorphic functions, as in [Gun90, p. 26]. Under these identifications, $\frac{\partial}{\partial x_k} \equiv \frac{\partial}{\partial z_k}$ and a vector (field)

$$v \equiv (v_k)_k \equiv (a_k + ib_k)_k \equiv \sum_k \left(a_k \frac{\partial}{\partial x_k} + b_k \frac{\partial}{\partial y_k} \right) \in \mathbb{C}^n \equiv T(\mathbb{C}^n)$$

is identified to the vector

$$\sum_k v_k \frac{\partial}{\partial z_k} \in T^{(1,0)}(\mathbb{C}^n) \cong T(\mathbb{C}^n) \cong \mathbb{C}^n.$$

See also [Nar85, pp. 74–80] and [Bog91].

Symmetric to $T^{(1,0)}$ there is the complex subspace

$$T^{(0,1)} \stackrel{\text{def}}{=} \{u \in \mathbb{C} \otimes T(\mathbb{C}^n) : J(u) = -i \cdot u\}$$

and the direct sum decomposition (over $\mathbb{C}$)

$$\mathbb{C} \otimes T(\mathbb{C}^n) = T^{(1,0)} \oplus_{\mathbb{C}} T^{(0,1)} .$$

The dual $J^* : T^*(\mathbb{C}^n) \to T^*(\mathbb{C}^n)$ of J is defined on the cotangent space by $\langle J^*(\phi), v\rangle = \langle \phi, J(v)\rangle$ for any $\phi \in T^*(\mathbb{C}^n)$ and $v \in T(\mathbb{C}^n)$. In particular, $J^*(dx_k) = -dy_k$ and $J^*(dy_k) = dx_k$. The operator J^* also extends to $\mathbb{C} \otimes T^*(\mathbb{C}^n)$ and induces a decomposition of $\mathbb{C} \otimes T^*(\mathbb{C}^n)$ into $(1,0)$ and $(0,1)$ forms. In particular, the exterior derivative $d\rho$ of a smooth function ρ decomposes as

$$d\rho = \partial\rho + \bar{\partial}\rho, \qquad \partial\rho \in T^{*(1,0)}, \qquad \bar{\partial}\rho \in T^{*(0,1)},$$

where

$$\partial\rho = \sum_{j=1}^{n} \frac{\partial \rho}{\partial z_k} \frac{\partial}{\partial z_k}$$

and

$$\bar{\partial}\rho = \sum_{j=1}^{n} \frac{\partial \rho}{\partial \bar{z}_k} \frac{\partial}{\partial \bar{z}_k} .$$

The Levi form

E.E. Levi [Lev10], [Lev11] proved that if a domain of holomorphy has smooth boundary, then a certain differential condition on the boundary, involving the complex tangent space, is satisfied. Let D be a **smoothly bounded domain** in $\mathbb{C}^n$, so that $D = \{z : q(z) < 0\}$ for a smooth real-valued function q whose gradient is different from 0 on the boundary of D (q is a **defining function** for D). E.E. Levi showed that if D is a domain of holomorphy then the complex Hessian

$$\mathsf{L} \stackrel{\text{def}}{=} \sum_{j,k=1}^{n} \frac{\partial \bar{\partial} q}{\partial z_j \partial \bar{z}_k}$$

is positive semi-definite on the complex tangent space $T^c_w(bD)$ for each point $w \in bD$. Moreover, he proved that if the **Levi form** L is positive definite on $T^c_w(bD)$, then a partial, local converse to his result holds. Smoothly bounded domains for which the Levi form is positive semi-definite on $T^c(bD)$ are called **Levi pseudoconvex**.

3.2. DOMAINS IN $\mathbb{C}^N$

We restrict our attention to smoothly bounded Levi pseudoconvex domains, since they are domains of holomorphy [Kra92a, p. 140, p. 211].

The decomposition of the tangent space $T(bD)$ into a complex tangent subspace $T^c(bD)$ and a complementary real one-dimensional subspace (the **missing direction**) also holds as before. Observe that the condition that a vector v belongs to $T(bD)$ can be written in terms of the defining function q: $v \in T(bD)$ if and only if $\langle dq, v \rangle = 0$, where $\langle \cdot, \cdot \rangle$ is the pairing between one forms and tangent vectors. A vector $v \in T(bD)$ belongs to $T^c(bD)$ if and only if $J(v)$ also belongs to $T(bD)$, i.e. that $\langle dq, J(v) \rangle = 0$; in particular, $\langle J^*(dq), v \rangle = 0$. The *distribution* $T^c(bD) \subset T(bD)$ is then defined as the set of vectors in $T(bD)$ on which the one-form

$$\beta \stackrel{\text{def}}{=} J^*(dq)$$

is zero. In the sequel, we write $v \in T(bD)$ to denote that v belongs to the tangent space to bD at some unspecified point, or that v is a section of the bundle $T(bD)$, i.e. a vector field, which is obtained from the extension of a fixed tangent vector at some fixed point, with the claim that the results do not depend on the choice of the extension.

Strongly Pseudoconvex Domains in $\mathbb{C}^n$

A smoothly bounded domain D in $\mathbb{C}^n$ for which the Levi form is positive definite on $T^c(bD)$ is called **strongly Levi pseudoconvex**. If D is strongly pseudoconvex, then the distribution $T^c(bD)$ is as far as possible from being integrable in the sense of Frobenius. Recall that, by the Frobenius theorem, a distribution on a 3-dimensional manifold, generated by a one-form β, is integrable exactly if $\beta \wedge d\beta = 0$. Now observe that Cartan's intrinsic definition of exterior derivative implies that, for each $v \in T^c(bD)$

$$\begin{aligned}\langle d\beta, v \wedge J(v) \rangle &= \langle v, \langle \beta, J(v) \rangle \rangle - \langle J(v), \langle \beta, v \rangle \rangle - \langle \beta, [v, J(v)] \rangle \\ &= -\langle \beta, [v, J(v)] \rangle\end{aligned}$$

where the brackets $[\cdot, \cdot]$ denote the Lie brackets of vector fields. A similar computation shows that

$$\langle \beta, [v, J(v)] \rangle = 2i \langle d\beta, V \wedge \overline{V} \rangle = 4\partial\overline{\partial} q \, V \wedge \overline{V} \qquad (3.7)$$

where

$$V = \frac{1}{2}(v - iJ(v)) \in T^{(1,0)}(bD) \stackrel{\text{def}}{=} T^{(1,0)}(\mathbb{C}^n) \cap [\mathbb{C} \otimes T(bD)].$$

Now the last term in (3.7) is exactly (a multiple of) the Levi form evaluated at the vector v. A **contact manifold** is a differentiable manifold of dimension $2k+1$, endowed with the choice of a codimension-one subbundle of its tangent bundle, defined by a one-form β, such that $\beta \wedge d\beta^k$ is a volume form; cf. [Arn74, Appendice 4H]. We conclude that, if D is strongly pseudoconvex, then $\langle d\beta, v \wedge J(v) \rangle \neq 0$ for $v \neq 0$, $v \in T^c(\mathrm{b}D)$. In particular, $\mathrm{b}D$ is a contact manifold.

The extension of Part 4 of Theorem 3.5 to strongly pseudoconvex domains and smoothly bounded domains in $\mathbb{C}^n$ is due to E.M. Stein [Ste72]; the relevant approach regions, also denoted by $\mathcal{A}_\alpha$, are modeled on the approach regions $\mathcal{A}_\alpha$ for the unit ball. This extension requires a method different from the one relying on equation (3.5), since the Poisson–Szegö is not known explicitly in this generality. The techniques that are used are based on the existence of embedded polydiscs inside the approach region $\mathcal{A}_\alpha$, and a repeated application, along each complex direction in the polydisc, of the sub-mean value property of holomorphic functions along analytic discs. See [Ste72], and [Hör67], [Bar78], [Lem80], [Kra92a, Lemma 8.6.10], [dB79], [Kra91]. A suitable version for convex domains of finite type in $\mathbb{C}^n$ [McN94] can be found in [DBF].

Pseudoconvex Domains of Finite Type in $\mathbb{C}^2$

The metric ρ defined for the unit ball in the previous section can also be interpreted in terms of the contact structure. Recall that the Lie bracket $[X,Y]$ of two vector fields defined on a differentiable manifold has the following property [Spi79, p. 220–225, Theorem 16]. Fix a point p in the manifold. Denote by $c(t)$ the point reached starting from p by first following the integral curve of X for time t; then following the one of Y for time t; then following the one for X backward for time t, and finally the one for Y backward for time t. Then

1. $c'(0) = 0$

2. $2[X,Y]_p \equiv c''(0)$, where $c''(o)$ is (the operator) defined by

$$c''(0)(f) \stackrel{\text{def}}{=} (f \circ c)''(0)$$

for each real-valued function f defined near p.

Now we restrict our attention to a domain D in $\mathbb{C}^2$ with smooth boundary. Let X and Y be a basis for the complex tangent space at each point of $\mathrm{b}D$. Fix a point $p \in \mathrm{b}D$. Consider the point $c(t)$ described above, obtained by following the flows of X and Y. An examination of the

3.2. DOMAINS IN $\mathbb{C}^N$

Taylor series expansion of the distance from $c(t)$ to p *along the missing direction* T reveals that the latter is of the order of t^2. In fact, the missing direction is precisely (a multiple of) $[X, Y]$. Observe that the exponent 2 is exactly the number of commutators required to span the direction along which we move. The meaning of the previous result is the following: If we follow the flows of X and Y for a time of the order of t, then we reach a point in the missing direction, whose distance from p is of the order of t^2. We can turn this fact backward, and *define* the anisotropic distance ρ in terms of flows along vector fields and their commutators, thus recapturing the metric ρ, or an equivalent version of ρ; see [NSW85].

The so-called egg domain

$$D = \{(z, \zeta) \in \mathbb{C}^2 : |z|^2 + |\zeta|^4 < 1\}$$

provides a model for the general class of pseudoconvex **finite type domains in** $\mathbb{C}^2$ (see [Koh72]). Let us use the coordinates

$$z = x + iy, \zeta = u + iv.$$

The vector fields

$$X_1 = 2(u^2 + v^2)u\frac{\partial}{\partial x} + -2(u^2 + v^2)v\frac{\partial}{\partial y} + (-x)\frac{\partial}{\partial u} + (y)\frac{\partial}{\partial v}$$

and

$$X_2 = 2v(u^2 + v^2)\frac{\partial}{\partial x} + 2u(u^2 + v^2)\frac{\partial}{\partial y} + (-y)\frac{\partial}{\partial u} - x\frac{\partial}{\partial v}$$

span the complex tangent space at each point, but the first commutator $[X_1, X_2]$ has a component equal to

$$u^2 + v^2$$

in the missing direction T, where

$$T = y\frac{\partial}{\partial x} - x\frac{\partial}{\partial y} + 2v(u^2 + v^2)\frac{\partial}{\partial u} - 2u(u^2 + v^2)\frac{\partial}{\partial v}.$$

In particular, on the circle $\zeta = 0$ in bD, the subbundle $T^c(bD)$ is not a contact subbundle. Another calculation reveals that

$$[X_1, [X_1, X_2]]$$

has a component equal to

$$16(-ux + vy)$$

in the T direction, while
$$[X_2, [X_1, X_2]]$$
has a component
$$-16(vx + uy)$$
in the missing direction T. If we differentiate one more time, we see that
$$[X_1, [X_1, [X_1, X_2]]] \text{ and } [X_2, [X_2, [X_1, X_2]]]$$
have components
$$16(-8u^4 - 16u^2v^2 - 8u^4 + x^2 + y^2)$$
in the T direction. In particular, the component is different from zero on bD. We see therefore that, by taking at most four commutators of X_1 and X_2, we do obtain the whole tangent space.

A metric is associated to any system of vector fields that have the property that a finite number of commutators span the tangent space at every point (the *Hörmander condition*). Carathéodory proved that, under these conditions, any two points can be joined by a piecewise smooth curve whose tangent belongs to the system of vector fields. A metric is then defined by giving weight δ^j to those directions obtained by commutators of j vector fields in the family. The precise definition, given below, can be formulated in various way, all leading to equivalent (quasi)-metrics; cf. [NSW81] and [NSW85].

Let D be a smoothly bounded pseudoconvex domain in $\mathbb{C}^2$. Let q be a defining function for D.

The shape of the balls and the associated approach regions for D turn out to depend on the size of the components along T of the commutators of X and Y. In fact, these sizes are involved in the definition of the "lambda function" $\Lambda(w, r)$ (defined below) where $w \in bD$ and $r > 0$. The Euclidean size of a ball $\mathbb{B}(w, r)$ in the metric ρ turns out to be of the order of r along the complex tangent directions, and of the order of $\Lambda(w, r)$ along the missing direction.

In the ball (and on a strongly pseudoconvex domain) the Λ-function is of the order of r^2. The computations made above for the egg domain show that the Lambda function $\Lambda(w, r)$ is of the order of r^4 for points in $\{\zeta = 0\} \cap bD$ (which have surface measure zero) and of the order of r^2 at each point w of bD outside the set $\{\zeta = 0\} \cap bD$, *but not uniformly in* w. In particular, if we approach the point $(1, 0) \in bD$ from the complex direction $(0, 0, 1, 0)$, along the curve
$$\gamma(t) \stackrel{\text{def}}{=} (\sqrt{1 - (t)^4}, 0, t, 0),$$

3.2. DOMAINS IN $\mathbb{C}^N$

we find that $\Lambda(\gamma(t), t)$ is of the order of t^4 as $t \to 0$.

The vector field
$$L = \frac{\partial q}{\partial \zeta}\frac{\partial}{\partial z} - \frac{\partial q}{\partial z}\frac{\partial}{\partial \zeta}$$
belongs to $[\mathbb{C} \otimes T(bD)] \cap T^{(1,0)}(\mathbb{C}^n)$. Therefore, it can be written as $L = \frac{1}{2}(X_1 - iX_2)$, where $X_2 = J(X_1)$. In fact, the vector fields X_1 and X_2 are those used in the computations made above. The vector T is a non-vanishing vector fields, and X_1, X_2, T span the whole tangent space $T(bD)$. For each k-tuple $I \equiv (I_1, I_2, \ldots, I_k)$ of integers belonging to the set $\{1, 2\}$, consider the commutator
$$[X_{I_k}, [X_{I_{k-1}}, \ldots, [X_{I_2}, X_{I_1}]\ldots]] = \lambda_I T + \alpha X_1 + \beta X_2 .$$
Thus λ_I gives the T component of the commutator of X_1, X_2 associated to I.

For each integer $\ell \geq 2$, let
$$\Lambda_\ell \stackrel{\text{def}}{=} \left(\sum_I |\lambda_I|^2\right)^{1/2} , \qquad (3.8)$$
where the sum is extended over all k-tuple's I such that $2 \leq k \leq \ell$.

A **point** $w \in bD$ is **of type** m if
$$\Lambda_2(w) = \Lambda_3(w) = \ldots \Lambda_{m-1}(w) = 0$$
but
$$\Lambda_m(w) \neq 0 .$$
The domain D is of type m if every point $w \in bD$ is of type at most m.

The Lambda function we alluded to above is defined by
$$\Lambda(w, r) \stackrel{\text{def}}{=} \sum_{j=2}^{m} \Lambda_j(w) r^j ,$$
where D is of type m.

The anisotropic metric ρ on bD is defined by letting
$$\rho(w, u) \stackrel{\text{def}}{=} \inf\{ \ r > 0 : \text{ there is a continuous piecewise smooth map}$$
$$\phi : [0, 1] \to bD \text{ such that}$$
$$\phi(0) = w, \phi(1) = u, \text{ and almost everywhere:}$$
$$\phi'(t) = \alpha_1(t)X_1 + \alpha_2(t)X_2 \text{ with}$$
$$\phi'(t) = |\alpha_1(t)| < r, |\alpha_2(t)| < r\}.$$

The approach regions for a finite-type domain are then defined in terms of the function

$$D(w,r) \stackrel{\text{def}}{=} \inf_{2 \leq k \leq m} \left(\frac{r}{\Lambda_k(w)} \right)^{1/k}.$$

Let D_1 be a small neighborhood of a point in the boundary of D such that for each point $z \in D_1$ there is one and only one point w in W, denoted by $\pi(z)$, such that $d(w,z) = d(z)$. Let

$$D(z) \stackrel{\text{def}}{=} \{D(\pi(z), d(z))\}$$

and define

$$\mathcal{A}_\alpha(w) \stackrel{\text{def}}{=} \{z \in D_1 : \pi(z) \in \mathbb{B}(w, \alpha \cdot D(z))\}. \qquad (3.9)$$

Theorem 3.6 *1. The space $W \stackrel{\text{def}}{=} (\mathrm{b}D, \nu, \rho)$ is a space of homogeneous type with respect to the metric ρ and the surface measure ν.*

2. The balls $\mathbb{B}(w, r)$ in the metric ρ have length of the order of $\Lambda(w, r)$ in the missing direction and length r in complex tangent directions.

3. The approach regions defined in (3.9) form an approach system: For each $\alpha > 0$ there are constants k_α and k'_α such that

$$k_\alpha \cdot \mathbb{B}(\pi(z), D(z)) \subset (\mathcal{A}_\alpha)^{\downarrow}(z) \subset k'_\alpha \cdot \mathbb{B}(\pi(z), D(z)).$$

4. Each bounded holomorphic function on D converges along $\mathcal{A}_\alpha$ on a subset of $\mathrm{b}D$ of full measure.

For a proof of this theorem, see [NSW81] and [NSW85]. The proof of Theorem 3.6 relies on the existence of embedded polydiscs $P_\varepsilon(z)$ in the approach regions $\mathcal{A}_\alpha(w)$, centered at points $z \in \mathcal{A}_\alpha(w)$, whose diameter is $\varepsilon d(z)$ in the complex normal direction (generated over $\mathbb{C}$ by N_u) and $D(z)$ in the complex tangent direction. The construction of such polydiscs is also encountered in [BDN91], [Cat89, pp. 432–439], [Gre92, pp. 258–260], [Koo93, pp. 10–22] and [McN94], [NRSW89], [Thi94]; cf. [DH92], [Dwi], [DH].

Part 4 of Theorem 3.6 also applies to holomorphic functions in the *Nevanlinna class* [NSW81], [NSW85]; cf. [Lem80], [Nef86], [Nef90].

3.3. TREES

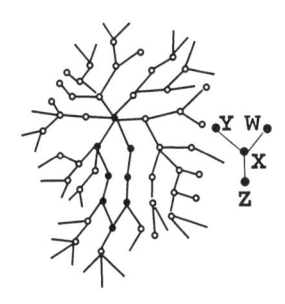

Figure 3.6: The graph on the left is not simply connected, since it has a loop. The vertex X has three adjacent vertices Y, Z and W.

3.3 Trees

A **tree** T is an infinite, connected graph that is locally finite, without loops, and where each vertex is adjacent to at least two other vertices. We write $x \sim y$ if the vertices $x, y \in T$ are adjacent. (See Fig. 3.6.) A basic reference is [Car72] (pp. 208–222).

A **path of length** N from $x \in T$ to $z \in T$ is a finite sequence $\{x(j)\}_{j=0}^{N}$ of successively adjacent vertices $x(j) \in T$, with $x(0) = x$ and $x(N) = z$. The **geodesic** from x to z is the shortest path from x to z. (See Fig. 3.7.)

An **edge** τ is an ordered pair $\tau = (x, y)$ of adjacent vertices. The collection of all edges is denoted by Λ. The identity $\tau = (b(\tau), e(\tau))$ defines the two projections $b, e : \Lambda \to T$. We let $\Lambda(x) = \{(x, y) : y \in T, y \sim x\}$, for $x \in T$.

The tree T is endowed with a **stochastic nearest neighbor operator** P on T, which is a function $P : \Lambda \to [0, 1]$ for which $\sum_{\tau \in \Lambda(x)} P(\tau) = 1$ for all $x \in T$. There are many such operators; we will assume that P satisfies the conditions stated below in (3.12). The operator P acts on functions defined on the tree as follows: $PF(x) = \sum_{\tau \in \Lambda(x)} P(\tau) F(e(\tau))$. The **Laplace operator** Δ induced by P is the operator $F \to \Delta(F)$ given by $\Delta F = PF - F$. A function F on T is **harmonic** if $\Delta F = 0$.

A tree T is a 0-*hyperbolic space in the sense of Gromov*, with respect to the natural **hyperbolic metric** $d_i(x, z)$ measuring the length of the geodesic from x to z. The associated **geometric boundary** bT of (T, d_i) can be described by fixing a **root** 0 in T, i.e. a *reference vertex*; then bT is the set of all **infinite geodesics** starting at 0 (i.e. sequences

$$\omega = \{\omega(j)\}_{j=0}^{\infty}$$

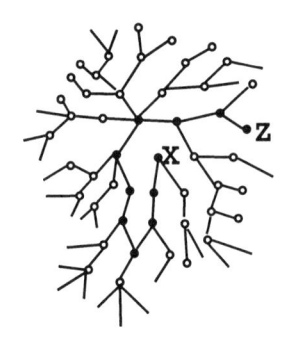

Figure 3.7: The geodesic from x to z.

of distinct, successively adjacent vertices $\omega(j) \in T$, with $\omega(0) = 0$); in particular, $d_i(0, \omega(j)) = j$ for all $\omega \in bT$ and $j = 0, 1, 2, \ldots$; see Chapters 6 and 7 in [GdlH90]. Given two distinct points ω, ω' in bT there is a unique double-sided sequence $\{x_j\}_{j \in \mathbb{Z}}$ of distinct, successively adjacent vertices in the tree, such that $x_0 = 0$, $\{x_{-j}\}_{j \geq 0} = \omega$ and $\{x_j\}_{j \geq 0} = \omega'$; this sequence $\{x_j\}_{j \in \mathbb{Z}}$ is called the **geodesic from** ω **to** ω'.

A **tree with root** $(T, 0)$ is a tree T in which a reference vertex $0 \in T$, the **root**, has been fixed.

Let $(T, 0)$ be a tree with root 0. The **partial ordering** $\leq$ induced on $T \cup bT$ by the root 0 is defined as follows: If $x, z \in T$ then

$$x \leq z$$

iff x belongs to the geodesic from 0 to z; if $x \in T$ and $\omega \in bT$ the

$$x \leq \omega$$

iff x belongs to the geodesic (from the root 0 to) ω (regarded as a subset of the tree). The root 0 is then a predecessor of all points in $T \cup bT$. If $x \leq z$, we say that z is a **descendant** of x, and that x is a **predecessor** of z. For every $x \in T$, $x \neq 0$, there exists exactly one **direct predecessor** x^- of x in this ordering; then x is a **direct descendant** of x^-. The set of direct descendants of a vertex x is denoted by $\lambda_+(x)$. Moreover, we write

$$|x|$$

to denote $d_i(x, 0)$, and we say that x belongs to the **generation** $|x|$. Observe that given two points x and y in $T \cup bT$, there is a unique vertex x ∧ y in T, called the **meet of** x **and** y, that is a predecessor of x and y and a descendant of any vertex that is a predecessor of x and y.

3.3. TREES

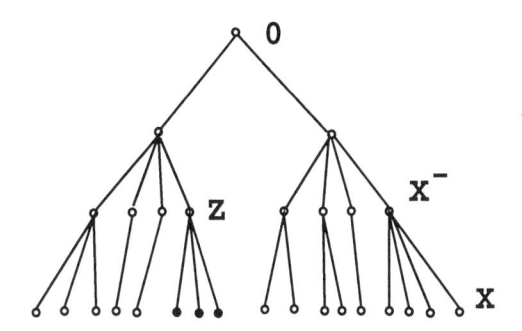

Figure 3.8: The root 0, the predecessor x^- of a vertex x, the descendants of the vertex z.

There is a natural metric d_e on $T \cup bT$, the *Euclidean metric*, for which $T \cup bT$ and bT are compact spaces. The **Euclidean metric** d_e is defined as
$$d_e(\mathsf{x}, \mathsf{y}) \stackrel{\text{def}}{=} e^{-|\mathsf{x} \wedge \mathsf{y}|} \quad \mathsf{x}, \mathsf{y} \in T \cup bT.$$
(See Fig. 3.10.) Moreover, we let
$$d_e(x, bT) \stackrel{\text{def}}{=} \inf\{d_e(x, \omega) : \omega \in bT\}$$
for any $x \in bT$. Observe that $d_e(x, bT) \equiv e^{-|x|}$. The **tent** $T(x)$ generated by x is the set $T(x) = \{z \in T : z \geq x\}$ of all descendants of x. (See Fig. 3.11.) The balls in bT relative to the metric d_e are the sets $E(x) \subset bT$, with $x \in T$, defined by
$$E(x) = \{\omega \in bT : x = \omega(j) \text{ for } j \equiv d_i(0,x) \}. \tag{3.10}$$
In fact, the open ball $\mathbb{B}(\omega, r) \subset bT$ of center ω and radius $r \in (0, e]$, in the metric d_e, can be written as
$$\mathbb{B}(\omega, r) = E(\omega(k+1)), \tag{3.11}$$
where $k \geq -1$ is an integer such that $e^{-k-1} < r \leq e^{-k}$; conversely, if x is a vertex whose hyperbolic distance from the origin is $k+1$, then $E(x) = \mathbb{B}(\omega, r)$ for all $\omega \in E(x)$. (See Fig. 3.11.) In particular, E is a map $T \to 2^{bT}$. Note that $T(x) = \{z \in T : E(z) \subset E(x)\}$.

The choice of a point $\omega_0 \in bT$, in the boundary of the tree, induces a partial ordering on T, also denoted by $\leq$, together with a partition of T into infinite subsets, called **generations**: $x \leq y$ (with respect to ω_0),

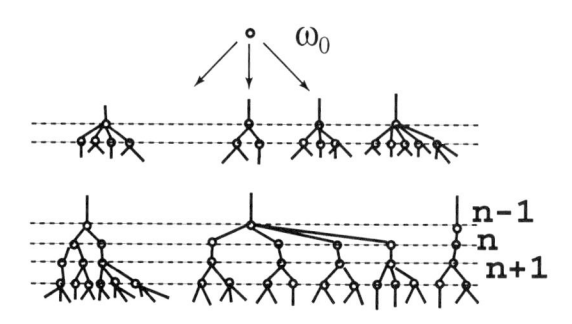

Figure 3.9: The "mythical ancestor" ω_0 does not belong to the tree. Generations are parametrized by an integer. Each generation contains infinitely many vertices.

if x belongs to the geodesic from y to ω_0. The definition of tent in this setting is the same as the one given above. The model case is given by the dyadic decomposition of $\mathbb{R}$. Once we decide which generation is the first, by the additional choice of a vertex $x_0 \in T$, then we can also define the generation $|x|$ of a vertex $x \in T$ relative to ω_0 and x_0. First, we say that x_0 belongs to the generation 0, and write $|x_0| = 0$. For any vertex $x \in T$, let us denote by y the unique point in the intersection of the geodesic from x_0 to ω_0 with the geodesic from x to ω_0, such that $d_i(x_0, y) + d_i(y, x)$ is minimal. We let
$$|x| \stackrel{\text{def}}{=} d_i(x, y) - d_i(y, x_0) \ .$$
The point ω_0 is then a predecessor of all points in the tree, but it does not belong to the tree[1]. For more details we refer the reader to [Car72, pp. 216–219]. A tree (T, ω_0) in which a point ω_0 in the boundary has been chosen is a **tree with root at infinity** and the point ω_0 is the **root at infinity** of (T, ω_0).

The operator P is **very regular** if *there is a constant* $\varepsilon_0 > 0$ *such that*
$$\varepsilon_0 \leq P(\tau) \leq \frac{1}{2} - \varepsilon_0 \qquad (3.12)$$
for all $\tau \in \Lambda$. We then say that T is a **very regular tree**.
The regularity condition (3.12) implies that the *random walk* $\{X_n\}_n$ starting at 0 and governed by P is *transient* [KPT88, p. 258]. The meaning of the transience is the following: With probability one, the random walk $\{X_n\}_n$ converges to a point X_∞ in the boundary bT. The number $1/2$ appears in a natural way, since we want to avoid the case in which the

---
[1] It is called a *mythical ancestor* in [Car72].

3.3. TREES

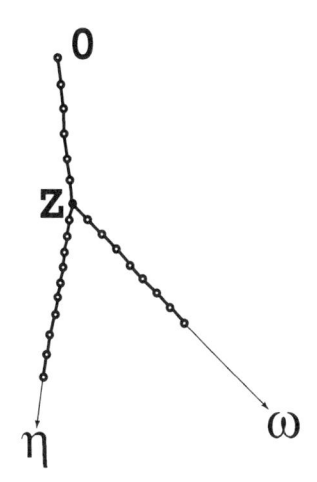

Figure 3.10: The vertex z is used to compute the Euclidean distance between ω and η.

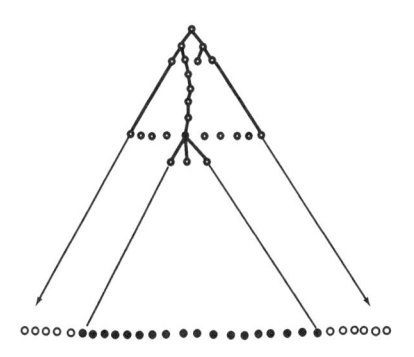

Figure 3.11: The tent $T(x)$ below x and the ball $E(x)$ in the boundary.

transition probabilities $\{p(x,y)\}_{y\sim x}$ at each vertex x are biased toward the predecessor x^- so as to force the random walk to visit the origin infinitely often with positive probability—if $p(x,x^-)$ is greater than the sum $\sum_{y\sim x, y\neq x} p(x,y)$ of the remaining probabilities, then $p(x,x^-) > 1/2$.

The distribution of the random variable X_∞ on $\mathrm{b}T$ is the **hitting distribution** ν of the random walk, i.e.

$$\nu(E(x)) \stackrel{\mathrm{def}}{=} \mathrm{Prob}\{X_\infty \in E(x)\}$$

for any $x \in T$; see [Car72, Théorème 2.1, p. 232; Théorème 3.1, p. 239; Section 3.4, p. 248].

As usual, for $E \subset \mathrm{b}T$, we write $\nu(E)$ as

$$|E|.$$

Theorem 3.7 *Let T be a tree with root, P a very regular operator on T, ν the hitting distribution of the random walk governed by P and starting at 0. Then $(\mathrm{b}T, \nu, d_e)$ is a space of homogeneous type.*

Theorem 3.7 is proved in [KPT88], where it is shown that the following uniformity holds: There is a constant ε, depending only on the ε_0 appearing in (3.12), such that for all vertices x different from the origin one has

$$0 < \varepsilon \leq \frac{|E(x)|}{|E(x^-)|} \leq 1 - \varepsilon < 1. \tag{3.13}$$

The doubling property then follows from (3.11). Observe that (3.13) is equivalent to

$$0 < \varepsilon \leq \frac{|E(x)|}{|E(x^-)|}.$$

A tree $(T, 0)$ with root is a space of approach to its boundary $\mathrm{b}T$, for the approach function d_e, the Euclidean metric. Consider the approach family for $(T, \mathrm{b}T)$ given by the family of *cones* $\Gamma_\alpha : \mathrm{b}T \to 2^T$:

$$\Gamma_\alpha(\omega) \stackrel{\mathrm{def}}{=} \{x \in T : d_i(x, \omega) \leq \alpha\},$$

where

$$d_i(x, \omega) \stackrel{\mathrm{def}}{=} \min_{j \in \mathbb{N}} d_i(x, \omega(j))$$

for $x \in T$ and $\omega \in \mathrm{b}T$. We will restrict the attention to integer values of α (or consider the integer part of α). Observe that these cones are cylinders in the hyperbolic metric of the tree.

3.3. TREES

The shadow of Γ_0 is precisely the map $E: T \to 2^{bT}$ defined in (3.10) by $E(x) = \{\omega \in bT : x \in \omega\}$. We then see that the tent $T(x)$ is precisely the tent over the ball $E(x)$ relative to Γ_0, i.e.

$$\Delta^{\Gamma_0}(E(x)) \equiv T(x)$$

for $x \in T$. The shadow $\Gamma_\alpha^\downarrow(x)$ is equal to the set $E(x[-\alpha])$:

$$\Gamma_\alpha^\downarrow(x) \equiv E(x[-\alpha]) \tag{3.14}$$

where $x[-\alpha]$ is the vertex at hyperbolic distance α from x, belonging to the geodesic from 0 to x, for $\alpha \leq |x|$ (if $\alpha > |x|$ then $x[-\alpha]$ is equal to the root 0). It follows that Γ_α is a natural approach family, and that $\{\Gamma_\alpha\}_\alpha$ is an approach system (of dilates of Γ_0).

Theorem 3.8 *Let $(T, 0)$ be a tree with root, endowed with a very regular operator P. Let ν be the hitting distribution on bT of the random walk governed by P and starting at the root. If F is a bounded harmonic function on the tree, then F converges along Γ_α on a subset of bT of full ν-measure.*

The convergence statement of Theorem 3.8 also holds for harmonic functions with h^p growth, $1 \leq p \leq \infty$. For a proof, see [KP86], and [Car72], [KPT88], [Tai75].

An approach family $\mathcal{L}$ is *exotic* at ω with respect to $\{\Gamma_\alpha\}_\alpha$ if $\mathcal{L}(\omega)$ contains a sequence x_n tending to ω but not contained in any cone $\Gamma_\alpha(\omega)$. Thus, an exotic approach family, near the point ω, lies outside any cone at ω, for almost every $\omega \in bT$. Observe that, on trees as well as on Euclidean half-spaces, the term "tangential" would be misleading, since these regions contain tangential sequences but do *not* contain tangential curves. More precisely, a **curve** s approaching a point $\omega_0 \in bT$ is a function $s : bT \setminus \{\omega_0\} \to T$ that is *continuous in the Euclidean metric d_e* and such that

$$s(\omega) \in \Gamma_0(\omega)$$

for all $\omega \neq \omega_0$, and

$$\lim_{\omega \xrightarrow{d_e} \omega_0} s(\omega) = \omega_0$$

in the Euclidean metric. The curve is **tangential** if

$$\lim_{\omega \xrightarrow{d_e} \omega_0} \frac{d_e(s(\omega), bT)}{d_e(s(\omega), \omega_0)} = 0 .$$

The convergence result of Theorem 3.8 fails for *tangential curves* on trees, as shown by [DB]—for general trees—and [SV94]—for homogeneous trees.

In Chapter 4 we construct regions of convergence that contain tangential sequences (as opposed to tangential curves). In particular, the distinction between tangential sequences and tangential curves, which is crucial in Euclidean half-spaces, is also meaningful *and* relevant in the discrete setting of trees, although one may be tempted to conclude that no kind of connectedness is required in the definition of a (tangential) curve on a tree, and define a (tangential) curve as a (tangential) sequence of points. More details are given in [DB].

Every approach family $\mathcal{L}$ induces a **maximal operator** $\mathcal{H}_\mathcal{L}$ on bT, in the following fashion. First we observe that every $f \in L^1(bT, \nu)$ extends to a function f_A on T by

$$f_A(x) = \frac{1}{|E(x)|} \int_{E(x)} f \, d\nu$$

for $x \in T$. Then, given a function $f \in \mathcal{L}^1(bT, \nu)$, we let

$$\mathcal{H}_\mathcal{L}(f)(\omega) \stackrel{\text{def}}{=} \sup_{x \in \mathcal{L}(\omega)} |f|_A(x)$$

for $\omega \in bT$.

The operator $\mathcal{H}_{\Gamma_0}$ is of weak type $(1,1)$, by Proposition 2.7, since Γ_0 is a natural approach family; we may also observe, more directly, that (3.11) implies that $\mathcal{H}_{\Gamma_0} f(\omega) = \sup_{r>0} \frac{1}{\nu(\mathbb{B}(\omega,r))} \int_{\mathbb{B}(\omega,r)} |f| \, d\nu$ and the conclusion follows from Theorem 1.13, as in the proof of Proposition 2.7. Recall that to every approach family $\mathcal{L}$ we associated an outer measure $\mathcal{L}^*$ on T:

$$\mathcal{L}^*(S) \stackrel{\text{def}}{=} \nu_e(\mathcal{L}^\downarrow(S))$$

for each $S \subset T$, where ν_e is the exterior measure generated by ν. Then $\mathcal{L}$ satisfies the Γ-**tent condition** if

$$\mathcal{L}^*(T(x)) \lesssim \nu(E(x))$$

for every $x \in T$. Recall that $T(x)$ is the Γ_0-tent over the ball $E(x)$.

The following lemma is a version of Theorem 2.14.

Lemma 3.9 *Let $\mathcal{L}$ be an approach family for (T, bT). The following are equivalent:*

1. *for all $x \in T$,*
$$\mathcal{L}^*(T(x)) \lesssim |E(x)|;$$

3.3. TREES

2. for all $f \in L^1(bT, \nu)$ and all $\lambda > 0$,
$$\nu_e\{\omega \in bT : \mathcal{H}_{\mathcal{L}}f(\omega) > \lambda\} \lesssim |\{\omega \in bT : \mathcal{H}_{\Gamma_0}f(\omega) > \lambda\}|;$$

3. for all $f \in L^1(bT, \nu)$ and all $\lambda > 0$,
$$\nu_e\{\omega \in bT : \mathcal{H}_{\mathcal{L}}f(\omega) > \lambda\} \lesssim \frac{1}{\lambda} \int_{bT} |f|\, d\nu.$$

Proof. We first prove that the Γ-tent condition (1) implies the distribution function inequality (2). The set
$$\{\mathcal{H}_{\Gamma_0}f > \lambda\} \stackrel{\text{def}}{=} \{\omega \in bT : \mathcal{H}_{\Gamma_0}f(\omega) > \lambda\}$$
is open, and therefore it admits a Whitney-type decomposition in balls, as in Lemma 1.15:
$$\{\mathcal{H}_{\Gamma_0}f > \lambda\} = \bigcup_j E(x_j). \tag{3.15}$$

While the general equation (3.15) would suffice, in our case, due to the special structure of (bT, ν, d_e), equation (3.15) specializes to a decomposition into disjoint balls, such that $E((x_j)^-) \not\subset \{\mathcal{H}_{\Gamma_0}f > \lambda\}$ for any j. Since any ball contained in $\{\mathcal{H}_{\Gamma_0}f > \lambda\}$ must also be contained in one of the balls $E(x_j)$, it follows that
$$\{|f|_A > \lambda\} \stackrel{\text{def}}{=} \{x \in T : |f|_A(x) > \lambda\} \subset \cup_j T(x_j).$$

Now it is clear that
$$\{\mathcal{H}_{\mathcal{L}}f > \lambda\} \stackrel{\text{def}}{=} \{\omega \in bT : \mathcal{H}_{\mathcal{L}}f(\omega) > \lambda\} = \mathcal{L}^{\downarrow}(\{|f|_A > \lambda\})$$

and therefore
$$\begin{aligned}
\nu_e\{\mathcal{H}_{\mathcal{L}}f > \lambda\} &= \nu_e \mathcal{L}^{\downarrow}\{|f|_A > \lambda\} \leq \nu_e \bigcup_j (\mathcal{L}^{\downarrow}(T(x_j))) \\
&\lesssim \sum_j |E(x_j)| = |\bigcup_j E(x_j)| \\
&= |\{\mathcal{H}_{\Gamma_0}f > \lambda\}|
\end{aligned}$$

using inequality (1) in the second inequality of the second line. Since the operator $\mathcal{H}_{\Gamma_0}$ is of weak-type $(1,1)$, we see immediately that the distribution function inequality (2) implies the weak-type inequality (3).

Finally, we show that the weak-type inequality (3) implies the Γ-tent condition (1). Let f be the characteristic function of the ball $E(x)$. If $\mathcal{L}(\omega) \cap T(x) \neq \emptyset$ then there is a point $z \in \mathcal{L}(\omega) \cap T(x)$. It follows that $E(z) \subset E(x)$ and therefore $|f|_A(z) = 1$. This means that $\mathcal{L}^{\downarrow}(T(x)) \subset \{\sup_{\mathcal{L}} |f|_A > 1/2\}$, and therefore

$$\nu_e(\mathcal{L}^{\downarrow}(T(x))) \leq \nu_e(\{\mathcal{H}_{\mathcal{L}} f > 1/2\}) \lesssim \int |f| \, d\nu = |E(x)|$$

where the constant in $\lesssim$ is the same as in (3), and therefore does not depend on x. This concludes the proof of the equivalence. **q.e.d.**

The Γ_0-completion of $\mathcal{L}$ is also called the **vertical completion** of $\mathcal{L}$, and denoted $\mathcal{L}_v$. It is defined by

$$\mathcal{L}_v(\omega) \stackrel{\text{def}}{=} \{x \in T : x \leq z \text{ for some } z \in \mathcal{L}(\omega)\}.$$

As in Theorem 2.14, we see that that $\mathcal{L}$ satisfies the Γ-tent condition if and only if $\mathcal{L}_v$ does. We will therefore assume, without loss of generality, that our approach family is vertical, i.e. that $\mathcal{L} = \mathcal{L}_v$.

The following lemma is a version of Theorem 2.9.

Lemma 3.10 *If the approach family $\mathcal{L}$ satisfies the Γ-tent condition, then the Poisson extension of a function $f \in L^1(bT)$ converges along $\mathcal{L}$ on a set of full measure in* bT.

Proof. The relevance of the maximal function $\mathcal{H}_{\mathcal{L}} f$ in the control of the boundary behaviour of harmonic h^p functions is based on the bound

$$\sup_{x \in \mathcal{L}(\omega)} |\mathcal{K}(f)(x)| \leq \mathcal{H}_{\mathcal{L}} f(\omega)$$

for all $\omega \in bT$. Here $\mathcal{K}(f)$ is the Poisson extension of $f \in L^1(bT)$, $\mathcal{K}$ the Poisson kernel. The bound follows from (i) the fact that the region is vertical, and (ii) the identity

$$\mathcal{K}(f)(x) = U(x, x^-) \mathcal{K}(f)(x^-) + (1 - U(x, x^-)) f_A(x), \quad (3.16)$$

valid for $x \neq 0$. For a definition of $U(\cdot, \cdot)$, $\mathcal{K}$ and $\mathcal{K}(f)$ and a proof, see [Car72, pp. 228–229, pp. 231–232] and [KPT88, p. 218].

The second ingredient is the following. If f is continuous on bT, then

$$\lim_{\substack{x \to \omega \\ de}} \mathcal{K}(f)(x) = f(\omega)$$

3.3. TREES

for all $\omega \in bT$ (unrestricted convergence). This follows from the fact that

$$\int_{bT} \mathcal{K}(x,\omega)\, d\nu(\omega) = 1$$

for all $x \in T$, and the previous identity (3.16), together with the estimate $U(x,y) \leq q$, for an absolute constant $q \in (0,1)$; see again [KPT88, p. 212]. The remaining part of the proof is standard, the only difference being in the use of the exterior measure ν_e in place of ν ([Sue92]; see Theorem 2.9 and [dG81]). **q.e.d.**

Part II
Exotic Approach Regions

Chapter 4

Approach Regions for Trees

In this chapter we prove the following

Theorem 4.1 [ADBU96] *Let T be a very regular tree. Then there exists an approach family $\mathcal{L}$ that is Nagel-Stein relative to the cones for the tree. In particular, harmonic functions with h^p growth do converge along $\mathcal{L}$ on a set of full measure of the boundary.*

4.1 The Dyadic Tree

We illustrate the technique in the simple case of a homogeneous tree T in which each vertex has three neighbors, and the harmonic functions are those relative to the isotropic kernel P, for which $P(x,y) = \frac{1}{3}$ whenever x and y are neighbors.

The geometric homogeneity gives to the tree a rich supply of isometries; moreover,

$$\nu(E(x)) = \nu(E(z)) \text{ whenever } d_i(x,0) = d_i(z,0) \qquad (4.1)$$

because of the homogeneity of the operator. In order to deal with a general non-homogeneous tree, with a non-homogeneous kernel, for which there are not many isometries, and (4.1) fails, we will need to recast the construction in different terms. We will recapture the lost symmetry by means of the notion of a *stopping time triangle* (see below), which amounts to a stopping time technique. This technique depends on the fact that the boundary of the tree is a space of homogeneous type.

It is enough to restrict our attention to the tent (the tree) below one of the three descendants of the root. In this way we study a tree in which every vertex has two descendants. This is called the **dyadic tree**.

Fix a **labeling** of the positively oriented edges, using -1 and 1 as labels. This means that for each vertex x, having descendants x_1, x_2, we associate -1 to the edge (x, x_1) and 1 to the edge (x, x_2) (many such choices are possible: Pick one).

For every point w in the boundary bT of T, we construct the region $\mathcal{L}(w)$, as follows. Recall that $\mathcal{L}(w)$ must be a subset of the tree whose closure, in the Euclidean metric d_e, contains w. The geodesic w is an infinite sequence of vertices $w(n)$. Consider the *dyadic block* of vertices $\{w(k) : k = 2^n, \ldots, 2^{n+1}\}$, and read the labels corresponding to the oriented edges $(w(k), w(k+1))$, where $k = 2^n, \ldots, 2^{n+1} - 1$, thus getting a finite sequence of 1's and -1's. We now change all the signs in this sequence, getting a new sequence, denoted by $s^*(w; n)$, and *apply* the sequence $s^*(w; n)$ to the vertex $w(2^n)$, getting a new vertex, denoted $l(w; n)$. More precisely, this means that we start from the vertex $w(2^n)$ and proceed downward along the path given by $s^*(w; n)$, according to the labeling. The region $\mathcal{L}(w)$ is defined as the union of the vertices $l(w; n)$ together with all their predecessors.

The approach family $\mathcal{L}$ constructed above is clearly exotic, since the hyperbolic distance from $l(w; n)$ to the geodesic w (regarded as a subset of T) is increasing with n; also, observe that $d_e(w, l(w; n))$ tends to 0 as $n \to \infty$.

The following result completes the proof of Theorem 4.1 in the homogeneous dyadic case:

Lemma 4.2 *The approach family constructed above satisfies the tent condition.*

Proof. For any point $x \in T$ there is a unique integer n such that $2^n \le |x| < 2^{n+1}$, and, corresponding to this n, a unique predecessor z_x of x, of generation 2^n. Consider the labels of the geodesic path from z_x to x. Now change all the signs in the labels appearing in the geodesic path from z_x to x, and follow the path from z_x corresponding to the changed signs. The endpoint of this path is a vertex x^* of the same generation as x. Moreover, $\mathcal{L}^{\downarrow}(E(x)) = E(x^*) \cup E(x)$. The result now follows from (4.1).
q.e.d.

4.2 The General Tree

In the case of the general tree there are two difficulties, which stem from a lack of uniformity. First, the tree does not have many isometries,

4.2. THE GENERAL TREE

since the number of descendants changes from point to point; second, for points x and y at the same distance from the origin, the measures $|E(x)|$ and $|E(y)|$ are not necessarily comparable, as shown by a random walk of Bernoulli type, in which the ratio will diverge (or tend to zero). It follows that the previous scheme will not work as it stands.

In order to describe our construction better, we need some new terminology.

Given a vertex $x \in T$, a **section** under x is a (finite) set of proper descendants of x such that every geodesic containing x contains one and only one point in the section. (See Figs. 4.1, 4.2, 4.3.)

Thus, a set $\mathcal{B}$ is a section under x iff

$$\text{the vertex } x \text{ has more than one direct descendant,} \tag{4.2}$$
$$\mathcal{B} \subset T(x) \setminus \{x\}, \tag{4.3}$$
$$E(x) = \bigcup_{z \in \mathcal{B}} E(z), \tag{4.4}$$
$$E(x) \cap E(y) = \emptyset \text{ for different } x, y \in \mathcal{B}. \tag{4.5}$$

Observe that a section under x is necessarily finite, since $E(x)$ is compact and the sets $\{E(y)\}_{y \in \mathcal{B}}$ are disjoint.

Given a vertex x and a section $\mathcal{B}$ under x, the set of points that lie between x and $\mathcal{B}$ is called a **triangle** with **origin** x and **basis** $\mathcal{B}$. Given a triangle Υ, we denote its origin by 0_Υ and its basis by $\mathcal{B}_\Upsilon$, so that

$$\Upsilon = \{z \in T(0_\Upsilon) : z \leq w \text{ for some } w \in \mathcal{B}_\Upsilon\}.$$

The **interior** Υ^0 of a triangle is the set $\Upsilon^0 = \Upsilon \setminus \mathcal{B}_\Upsilon \setminus \{0_\Upsilon\}$.
Every triangle Υ defines a function (denoted by the same symbol Υ)

$$\Upsilon : T \cup bT \to 2^T,$$

as follows: For $w \in T \cup bT$, $\Upsilon(w)$ is actually a subset of the basis $\mathcal{B}_\Upsilon$ of Υ, and is called the **trace** of w on Υ. For a vertex $z \in T$, we let

$$\Upsilon(z) \stackrel{\text{def}}{=} T(z) \cap \mathcal{B}_\Upsilon.$$

Observe that $\Upsilon(z) \neq \emptyset$ only if $z \in \Upsilon$ or if z is a predecessor of 0_Υ. We let

$$\Upsilon(\omega) \stackrel{\text{def}}{=} \{z \in \mathcal{B}_\Upsilon : \omega \in E(z)\} \tag{4.6}$$

for every $\omega \in bT$. The set $\Upsilon(\omega)$ is either empty or consists of a single element, also denoted by $\Upsilon(\omega)$. In fact, $\Upsilon(\omega) \neq \emptyset$ only if $\omega \in E(0_\Upsilon)$. (See Figs. 4.4, 4.5.)

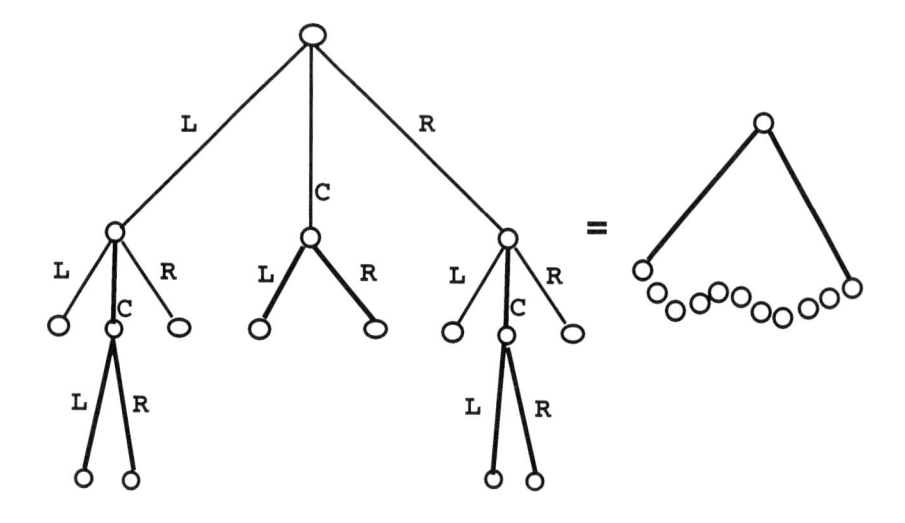

Figure 4.1: A section and the corresponding triangle; on the right the schematic picture that suggests the terminology. Observe that by repetition of the given triangle we get one of the three identical pieces that form the boundary of the snowflake. The labeling corresponds to the choices available at each stage (Left, Center, Right).

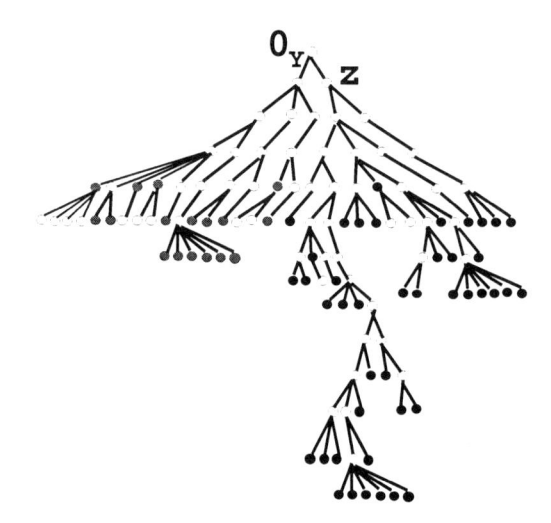

Figure 4.2: Another example of a triangle Υ in a tree, also showing the trace of a vertex z on Υ.

4.2. THE GENERAL TREE

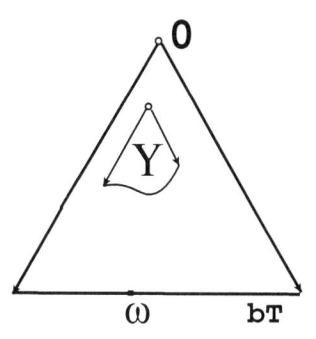

Figure 4.3: A "low resolution picture" of a triangle in a tree.

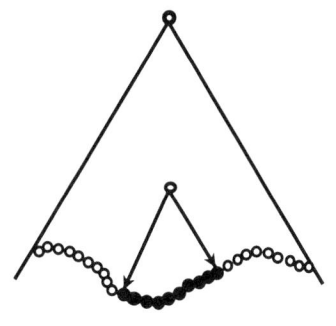

Figure 4.4: A schematic picture of the trace of a vertex on a triangle.

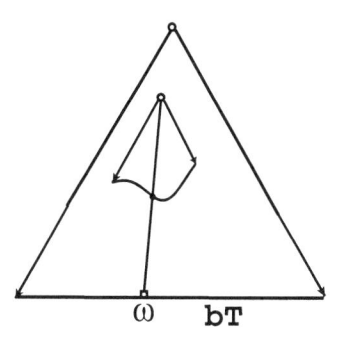

Figure 4.5: The trace of a point in bT on a triangle.

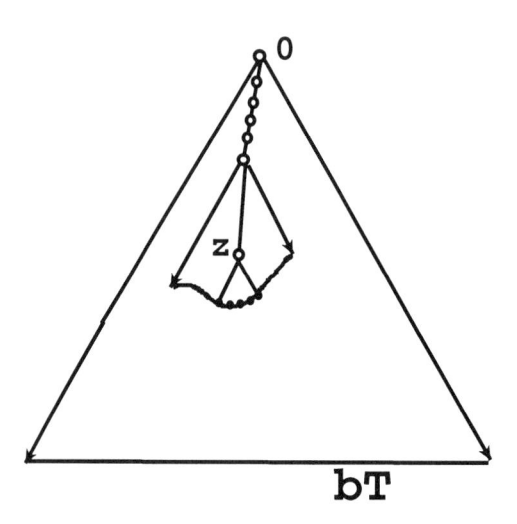

Figure 4.6: The trace of a vertex on a triangle.

The **height** $h(\Upsilon)$ of a triangle Υ is the lowest hyperbolic distance from its origin to its basis, i.e.

$$h(\Upsilon) \stackrel{\text{def}}{=} \min\{d_i(y, 0_\Upsilon) : y \in \mathcal{B}_\Upsilon\}. \tag{4.7}$$

We say that two triangles Υ_1 and Υ_2 are **separated** if they are either disjoint or the origin of one of them belongs to the basis of the other, and, therefore, they don't intersect in other points.

A **twisted triangle** ξ is a pair $\xi = (\Upsilon, \sigma)$, where $\Upsilon = \Upsilon_\xi$ is a triangle and $\sigma = \sigma_\xi$ is a map $\sigma : \mathcal{B}_\Upsilon \to \mathcal{B}_\Upsilon$, called a **twist** of Υ.

To each twisted triangle $\xi = (\Upsilon, \sigma)$ we associate an approach family, denoted by $\mathcal{L}_\xi$ and called the **approach family spanned by the twisted triangle** ξ, by means of the following: First define

$$l_\xi(\omega) = \{\omega(n)\}_{n \geq 0} \cup \{\sigma(x) : x \in \Upsilon(\omega)\}.$$

Then let $\mathcal{L}_\xi(\omega)$ be the vertical completion of $l_\xi(\omega)$, i. e.

$$\mathcal{L}_\xi(\omega) = \{x \in T : x \leq z \text{ for some } z \in l_\xi(\omega)\}.$$

Observe that, if ω does not belong to $E(0_\Upsilon)$, then $\mathcal{L}_\xi(\omega)$ contains only the geodesic $\{\omega(n)\}_{n \in \mathbb{N}}$. We already observed that, if $\omega \in E(0_\Upsilon)$, then there is precisely one element $\Upsilon(\omega) \in \mathcal{B}_\Upsilon$ such that $\omega \in E(\Upsilon(\omega))$, and $\mathcal{L}_\xi(\omega)$ contains, together with the geodesic ω, the finite geodesic from 0_Υ

4.2. THE GENERAL TREE

to $\sigma(\Upsilon(\omega))$. We denote the finite geodesic from 0_Υ to $\sigma(\Upsilon(\omega))$ by $b_\xi(\omega)$. The hyperbolic length of a branch $b_\xi(\omega)$ is no less than the height $h(\Upsilon_\xi)$.

If $\xi = (\Upsilon, \sigma)$ is a twisted triangle and $\mathcal{L}_\xi$ the associated approach family, then for each $x \in T$ there is a subset $\xi_*(x) \subset \mathcal{B}_\Upsilon$ such that

$$(\mathcal{L}_\xi)^\downarrow(x) = \bigcup_{z \in \xi_*(x)} E(z) \cup E(x). \tag{4.8}$$

If $x \notin \Upsilon$ then $\xi_*(x) = \emptyset$. If $x \in \Upsilon$ then $\xi_*(x) = \sigma^{-1}(\Upsilon(x))$. Thus we get a map

$$\xi_* : T \to 2^T.$$

The map ξ_* completely determines the approach family $\mathcal{L}_\xi$ via the duality (4.8) between $\mathcal{L}_\xi$ and ξ_*.

These observations say that the area of influence of $(\mathcal{L}_\xi)$ is precisely equal to Υ. Now, given a sequence $\{\xi_n\}_{n \in \mathbb{N}} = \{(\Upsilon_n, \sigma_n)\}_{n \in \mathbb{N}}$ of twisted triangles, we consider the associated approach families $\mathcal{L}_{\xi_n}$ and let $\mathcal{L}(\omega) = \bigcup_n \mathcal{L}_{\xi_n}(\omega)$. We will see that, under certain conditions on the sequence $\{\xi_n\}_{n \in \mathbb{N}}$, the approach family $\mathcal{L}$ turns out to satisfy the tent condition, and is also exotic. In the meantime, let us observe that, if $\{\mathcal{L}_n\}_{n=1}^\infty$ is a sequence of approach families, and if $\mathcal{L}$ is defined by

$$\mathcal{L}(\omega) = \bigcup_{n=1}^\infty \mathcal{L}_n(\omega),$$

then

$$\mathcal{L}^\downarrow(x) = \bigcup_{n=1}^\infty (\mathcal{L}_n)^\downarrow(x)$$

for all $x \in T$.

In order to construct the sequence $\{\xi_n\}_{n \in \mathbb{N}}$, we need a few more definitions.

If x is a vertex of T, and $y \in T(x)$ is a descendant of x, then the **relative weight of y with respect to x**, $\|y\|_x$, is defined to be

$$\|y\|_x := \frac{|E(y)|}{|E(x)|}, \tag{4.9}$$

so that, in particular, $\|y\| := \|y\|_0 = |E(y)|$.

If x is a vertex of T having more than one direct descendant, and j is a large positive integer, we say that $y \in T$ is a **stopping point under x of time j**, if

$$y \in T(x) \text{ and } \|y\|_x \le \varepsilon^j < \|y^-\|_x, \tag{4.10}$$

where y^- is the predecessor of y. Observe that (3.13) implies that $||y||_x > \varepsilon^{j+1}$ for any stopping point y under x of time j.

Let $S_{(x,j)}$ be the set of all stopping points under x of time j. The set $S_{(x,j)}$ is called the **stopping-time section under** x **at time** j. This terminology is justified by the following

Claim 4.3 *The set $S_{(x,j)}$ is a section under x.*

Proof. Conditions (4.3) and (4.5) are immediate consequences of the definition, while (4.4) follows from the fact that $||\omega(n)|| \to 0$ as $n \to \infty$ for any $\omega \in bT$. q.e.d.

The triangle corresponding to the section $S_{(x,j)}$ is called a **stopping-time triangle** and is denoted by $\Upsilon_{(x,j)}$.

It follows from (3.12) and (3.13) that $\frac{1}{1-\varepsilon} \leq \#\lambda_+(x) \leq \frac{1}{\varepsilon}$ (in fact $\#\lambda_+(x) \geq 2$), where $\lambda_+(x)$ is the set of direct descendants of the vertex x, and $\#A$ is the cardinality of a finite set A. Now consider a stopping-time triangle $\Upsilon_{(x,j)}$, and observe that (3.13) implies the bound

$$\frac{\varepsilon^2}{1-\varepsilon} \leq \frac{\#\Upsilon_{(x,j)}(z)}{\#\Upsilon_{(x,j)}(y)} \leq \frac{1-\varepsilon}{\varepsilon^2} \qquad z, y \in \lambda_+(x) \qquad (4.11)$$

for the cardinalities of the traces of the triangle on direct descendants of its origin x. The stopping condition implies that

$$\varepsilon \leq \frac{||y||_x}{||z||_x} = \frac{||y||}{||z||} \leq \frac{1}{\varepsilon} \qquad (4.12)$$

for any $y, z \in S_{(x,j)}$.

Definition 4.4 *A map $\phi : A \to B$ between finite sets A and B, of cardinalities $\#A$ and $\#B$ respectively, is called* balanced *if*

$$\#\phi^{-1}(b) \leq 1 + \frac{\#A}{\#B} \qquad (4.13)$$

for all $b \in B$. A positive number c is called a bound *for the map ϕ if $\#\phi^{-1}(b) \leq c$ for all $b \in B$.*

There are many balanced maps between any two finite sets. A twist σ of a triangle Υ is **regular** if there is a map

$$g = g_\sigma : \lambda_+(0_\Upsilon) \to \lambda_+(0_\Upsilon)$$

from the set $\lambda_+(0_\Upsilon)$ of all direct descendants of 0_Υ into itself, such that

4.2. THE GENERAL TREE

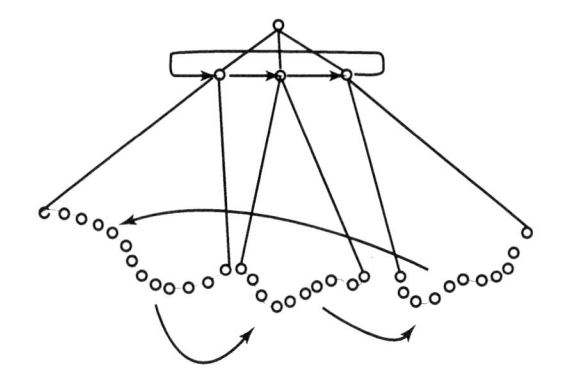

Figure 4.7: A regular twist. The origin of the triangle has three direct descendants.

1. g is one-to-one and onto, and *fixed point free*,

2. σ restricts to a balanced map $\Upsilon(y) \to \Upsilon(g(y))$, for all $y \in \lambda_+(0_\Upsilon)$.

Every triangle has many regular twists.

Proof of Theorem 4.1. Let us construct an appropriate sequence $\{\xi_n\}_{n=1}^\infty = \{(\Upsilon_n, \sigma_n)\}_{n=1}^\infty$ of twisted triangles. We choose Υ_n to be $\Upsilon_{(x_n, \ell_n)}$, where $\ell_n = 2^n$ and the sequence $\{x_n\}_{n=1}^\infty$ is selected in such a way that

(i) Υ_n is separated from Υ_m for $n \neq m$,

(ii) each $w \in bT$ belongs to $E(x_n)$ for infinitely many values of n.

(Define Υ_1 by choosing $x_1 = 0$; then form another piece of the sequence, say $x_2, x_3, \ldots, x_k$, by taking all the elements of $\mathcal{B}_1 = \mathcal{B}_{\Upsilon_1}$, where $k - 1 = \#(\mathcal{B}_1)$; proceed in this way so that whenever $x := x_n$ is an element of the sequence, and $\mathcal{B}_n := \mathcal{B}_{\Upsilon_n}$, then all points of $\mathcal{B}_n$ belong to the sequence.)

For each triangle Υ_n, choose a regular twist σ_n. Now we consider the associated approach families $\mathcal{L}_{\xi_n}$ and define

$$\mathcal{L}(w) = \bigcup_n \mathcal{L}_{\xi_n}(w).$$

We are now ready to prove the following claim.

Claim 4.5 *The approach family $\mathcal{L}$ is Γ-exotic.*

Proof. Let $w \in bT$. By (ii) above, w belongs to $E(x_n)$ for infinitely many values of n, say $w \in E(x_{n_j})$ for some sequence $n_1 < n_2 < \ldots < n_j < n_{j+1} < \ldots$, and therefore $\ell_{n_j} \to \infty$. Correspondingly, the heights $h(\Upsilon_{n_j})$ of the triangles Υ_{n_j} will tend to infinity, since $h(\Upsilon_{(x,\ell)}) \geq \ell$, again by (3.13). Since the twists are regular, one has

$$d_i(w, \sigma_{n_j}(\Upsilon_{n_j}(w))) = d_i(0_\Upsilon, \sigma_{n_j}(\Upsilon_{n_j}(w))) \geq h(\Upsilon_{n_j}),$$

and this shows that $\mathcal{L}(w)$ is exotic, since, as observed before, the right hand side of the previous line tends to infinity. q.e.d.

The following claim concludes the proof of Theorem 4.1.

Claim 4.6 *The approach family $\mathcal{L}$ satisfies the Γ-tent condition.*

Proof. We must show that

$$|\mathcal{L}^\downarrow(x)| \lesssim |E(x)|.$$

Let $x \in T$. Since the triangles are separated, only one of the following cases can occur:

(a) There are two integers n, m such that $x \in \mathcal{B}_{\Upsilon_n}$ and $x = 0_{\Upsilon_m}$;

(b) there is some integer n such that $x \in (\Upsilon_n)^0$.

Let us first examine case (a). Since x does not belong to the interior of any other triangle, it follows that $\mathcal{L}^\downarrow(x) = \bigcup_{(\xi_n)_*(x)} E(y) \cup E(x)$, where $(\xi_n)_*(x) = \sigma_n^{-1}(\Upsilon_n(x)) = \sigma_n^{-1}(x) = \{y \in \mathcal{B}_{\Upsilon_n} : \sigma_n(y) = x\}$. Therefore

$$|\mathcal{L}^\downarrow(x)| = ||x|| + \sum_{y \in \sigma_n^{-1}(x)} ||y|| \leq ||x|| + \frac{1}{\varepsilon}||x|| \cdot (\#\sigma_n^{-1}(x)) \leq c_\varepsilon ||x||$$

follows from (4.12), (4.11) and the definition (4.13) of a balanced map, with $c_\varepsilon = 1 + \frac{1}{\varepsilon}(1 + \frac{1-\varepsilon}{\varepsilon^2})$.

We finally examine case (b). Consider the set $\Upsilon_n(x)$, and observe that

$$||x|| = \sum_{y \in \Upsilon_n(x)} ||y||, \qquad (4.14)$$

since Υ_n is a triangle and ν is a measure.

Consider also the corresponding inverse image $\sigma_n^{-1}(\Upsilon_n(x))$.

As observed before, $(\xi_n)_*(x) = \sigma_n^{-1}(\Upsilon_n(x))$. Moreover, since the various triangles are separated, one has the following localization:

$$\mathcal{L}^\downarrow(x) = \bigcup_{z \in (\xi_n)_*(x)} E(z) \cup E(x). \qquad (4.15)$$

4.2. THE GENERAL TREE

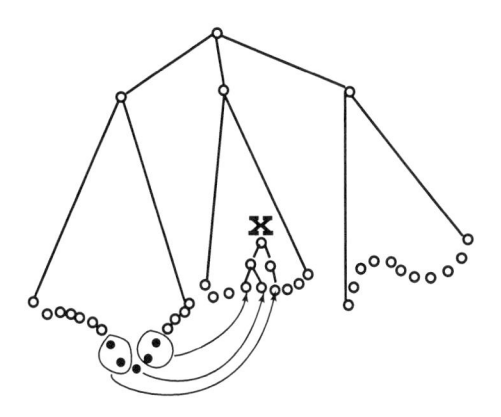

Figure 4.8: Case (b). The black vertices represent $(\xi_n)_*(x)$.

(See Fig. 4.8.)

Now (4.12), (4.11), (4.13), (4.15) and (4.14) imply that

$$\begin{aligned}\nu(\mathcal{L}^{\downarrow}(x)) &= ||x|| + \sum_{y \in \Upsilon_n(x)} \sum_{\sigma_n(z)=y} ||z|| \leq ||x|| + \sum_{y \in \Upsilon_n(x)} \sum_{\sigma_n(z)=y} \frac{1}{\varepsilon} ||y|| \\ &\leq ||x|| + \frac{1}{\varepsilon} \sum_{y \in \Upsilon_n(x)} ||y|| \cdot (\#\sigma_n^{-1}(y)) \leq c_\varepsilon \, ||x|| = c_\varepsilon \, \nu(E(x)) \, .\end{aligned}$$

q.e.d.

Chapter 5

Embedded Trees

5.1 The Unit Disc

In this section we illustrate our main theorem, Theorem 5.12, in the simple, crucial case of the unit disc, showing that there is a natural way to embed the tree in the unit disc, and to *transplant* the Nagel–Stein approach family constructed for the tree to a Nagel–Stein approach family for the unit disc. This is made more precise in the following Theorem 5.1, which is a special case of Theorem 5.12.

Theorem 5.1 *Let $\mathcal{L}$ be the Nagel–Stein approach family constructed for the dyadic tree in the previous chapter. Then it is possible to embed $\mathcal{L}$ inside the unit disc, so that the embedded approach family is Nagel–Stein relative to the nontangential approach regions for the unit disc.*

For convenience we work in the Euclidean half-plane. We can localize the construction to the interval $[0, 1]$.

Let T be the set of all dyadic intervals contained in $[0, 1]$ —recall that an interval in $[0, 1]$ is called **dyadic** if it is of the form $(\frac{j}{2^k}, \frac{j+1}{2^k})$ where j, k are positive integers.

The set T is naturally embedded in the upper half-plane, as follows. We associate, to each dyadic interval I, the point $z(I)$ in the upper half-plane, belonging to the normal segment issuing from the center of I, at height equal to the length of I. (See Figure 5.1.)

The set T is naturally endowed with the structure of the dyadic tree, in which the origin 0 is the interval $(0, 1)$. Each vertex in T has two descendants. For example, $(0, 1)$ has $(0, 1/2)$ and $(1/2, 1)$ as descendants; $(0, 1/2)$ has $(0, 1/4)$ and $(1/4, 1/2)$ as descendants, and so on.

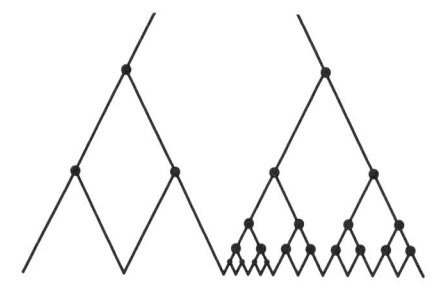

Figure 5.1: Part of the embedded tree.

With the exception of dyadic rationals, each point

$$\omega \in [0,1]$$

is uniquely identified by the sequence

$$\{I_n(\omega)\}_{n=0}^{\infty}$$

of dyadic intervals containing it, where the length of $I_n(\omega)$ is 2^{-n}. The sequence $\{I_n(\omega)\}_{n=0}^{\infty}$ is an infinite geodesic in the tree T. Conversely, every infinite geodesic in T is an infinite sequence $\{I_n(\omega)\}_{n=0}^{\infty}$ of dyadic intervals, and therefore it corresponds to a well determined point in $[0,1]$. This correspondence between $[0,1]$ and the boundary bT of T is one-to-one and onto, with the exception of the set of measure zero of dyadic rationals. The correspondence is also measure-preserving, if we endow each space with its natural measure: the interval $[0,1]$ with the Lebesgue measure, the boundary bT of T with the measure ν for which $\nu(E(x)) = 2^{-n}$ if $d_i(x,0) = n$.

Therefore the tree T is embedded in the half-plane. Consequently, a vertex $z \in T$ represents, at the same time, a vertex in the dyadic tree, the corresponding dyadic interval in $[0,1]$, and the embedded point in the upper half-plane.

Corresponding to the embedding described above, there is a natural labeling of the positively oriented edges, in which we associate -1 to the edge pointing to the "descendant on the left", and 1 to the edge pointing to the "descendant on the right".

Each number $\omega \in [0,1]$ has the dyadic expansion

$$\omega = \sum_{j=0}^{\infty} a_j \beta_j(\omega), \qquad (5.1)$$

5.1. THE UNIT DISC

where $a_j = 2^{-j-1}$ and β_j are the Rademacher functions, namely $\beta_0 \equiv 1$, $\beta_1 = -1$ on $(0, 1/2)$, $\beta_1 = 1$ on $(1/2, 1)$, and so on, so that $\beta_j(\omega) = -\text{sign}(\sin(2^{j-1}2\pi\omega))$ for $j = 1, 2, \ldots$ (note that we have changed signs with respect to the usual convention).

The dyadic expansion (5.1) is another expression of the various identifications made above. In particular, it is coherent with the choice of the labeling of T. Thus, given any $\omega \in [0, 1]$, the sequence $\{I_j(\omega)\}_j$ of dyadic intervals containing ω can be identified with the geodesic $\{\omega(j)\}_j$ in the tree, which is a sequence of points in the upper half-plane, given by the coordinates:

$$\omega(j) = \left(\sum_{k=0}^{j} a_k \beta_k(\omega), 2^{-j} \right).$$

We group the terms of (5.1) into dyadic blocks:

$$\Delta_p(\omega) = \sum_{j=1+2^p}^{2^{p+1}} a_j \beta_j(\omega)$$

for $p = 0, 1, 2, \ldots$, with $\Delta_{-1}(\omega) = a_0 \beta_0(\omega) + a_1 \beta_1(\omega)$. We then consider the point

$$e_p(\omega) = \left(\sum_{j=-1}^{p-1} \Delta_j(\omega) \right) - \Delta_p(\omega),$$

which belongs to $[0, 1]$. Finally we consider the point

$$l_p(\omega) = (e_p(\omega), 2^{-2^{p+1}})$$

belonging to the upper half-plane and corresponding to one of the vertices of the embedded tree. The reader should verify that the points $l_p(\omega)$ are the embedded images of the points $l(\tilde{\omega}; p)$ defined in Section 2.1, where $\tilde{\omega}$ is the point in the dyadic tree corresponding to ω in the identification between $[0, 1]$ and the boundary of the dyadic tree.

Recall that, given a point $(\omega, y) \in \mathbb{R} \times [0, \infty)$ in the closed upper half-plane, the *Euclidean cone* of aperture one, with vertex in (ω, y), is the set

$$\Gamma(\omega, y) \stackrel{\text{def}}{=} \{(v, t) : v \in \mathbb{R},\ t \geq 0,\ |v - \omega| \leq t - y\}.$$

We let $\Gamma(\omega) \equiv \Gamma(\omega, 0)$.

For each p and ω, let $L_p(\omega) \stackrel{\text{def}}{=} \Gamma(l_p(\omega))$ denote the cone, of aperture one, with vertex in the point $l_p(\omega)$, and define $\mathcal{L}(\omega) = \bigcup_{p=1}^{\infty} L_p(\omega) \cup \bigcup_j \Gamma(\omega(j))$.

The following two lemmas complete the proof of Theorem 5.1.

Lemma 5.2 *For almost every $\omega \in [0,1]$, the region $\mathcal{L}(\omega)$ contains a sequence approaching ω tangentially.*

Lemma 5.3 *The approach region $\mathcal{L}$ satisfies the tent condition for the upper half-plane.*

Proof of Lemma 5.2. Observe that $l_p(\omega) \notin \Gamma_\alpha(\omega)$ if and only if $|\omega - e_p(\omega)| > \alpha 2^{-2^{p+1}}$. Thus, the sequence $\{l_j(\omega)\}_{j \in \mathbb{N}}$ contains a tangential subsequence if and only if, for all $\alpha > 0$, we have $2^{2^{p+1}} |\omega - e_p(\omega)| > \alpha$ for infinitely many p's, i.e.

$$\sup_{p \geq 0} 2^{2^{p+1}} |\omega - e_p(\omega)| = +\infty . \tag{5.2}$$

Now,

$$|\omega - e_p(\omega)| = \left| 2\Delta_p(\omega) + \sum_{j=1+2^{p+1}}^{\infty} a_j \beta_j(\omega) \right|$$

and

$$\left| \sum_{j=1+2^{p+1}}^{\infty} a_j \beta_j(\omega) \right| \leq \frac{1}{2} 2^{-2^{p+1}}$$

imply that (5.2) is equivalent to

$$\sup_{p \geq 0} 2^{2^{p+1}} |\Delta_p(\omega)| = \infty .$$

It follows that the set

$$E \overset{\text{def}}{=} \{\omega \in bT : \text{ the sequence } \{l_j(\omega)\}_j \text{ is not exotic }\}$$

is the union of the countable family $\{E_n\}_{n=1}^{\infty}$, where

$$E_n \overset{\text{def}}{=} \{\omega : \sup_p 2^{2^{p+1}} |\Delta_p(\omega)| \leq n\} .$$

Therefore it suffices to show that $|E_n| = 0$ for all $n \in \mathbb{N}$—here vertical bars denote Lebesgue measure.

Fix $n \in \mathbb{N}$ and let $\omega \in E_n$. We can assume that ω is not a dyadic rational. Let B be the dyadic interval of length 2^{-2^k} that contains ω, where $k \in \mathbb{N}$. Since the center of B is

$$\gamma_B = \sum_{j=0}^{2^k} a_j \beta_j(\omega) ,$$

5.1. THE UNIT DISC

it follows that

$$|\omega - \gamma_B| \le \sum_{j=k}^{\infty} |\Delta_j(\omega)| \le n \sum_{j=k}^{\infty} 2^{-2^{j+1}} \le 2n2^{-2^{k+1}}$$

and therefore

$$|E_n \cap B| \le 4n2^{-2^{k+1}}.$$

Summing this relation over all dyadic intervals B of length 2^{-2^k}, we see that

$$|E_n| \le 4n2^{-2^k}$$

for all $k > 0$. Now let $k \to \infty$ to get $|E_n| = 0$. **q.e.d.**

Proof of Lemma 5.3. For each interval I, let $\Delta(I)$ be the tent over I, i.e.

$$\Delta_I = \{z \in [0,1] \times (0,\infty) : \Gamma^\downarrow(z) \subset I\}$$

where $\Gamma^\downarrow(z) = \{\omega \in [0,1] : z \in \Gamma(\omega)\}$. We need to show that

$$|\mathcal{L}^\downarrow(\Delta(I))| \le c\,|I| \tag{5.3}$$

for some constant c independent of the interval I, where, again,

$$\mathcal{L}^\downarrow(\Delta(I)) = \{\omega \in [0,1] : \mathcal{L}(\omega) \cap \Delta(I) \ne \emptyset\}.$$

In fact, (5.3) is the tent condition for the Euclidean half-plane.

Now, (5.3) holds when I is a dyadic interval, with $c = 2$. This is a crucial fact, but we leave its verification to the reader, since the verification is basically identical to that of Lemma 4.2.

The general case is tamed by way of a *decomposition of maximal type*, (a special case of Theorem 5.10) as follows; cf. Section 5.3. Every interval I in $[0,1]$ admits a maximal decomposition

$$I = \bigcup_j I_j \cup N$$

into disjoint dyadic intervals I_j, where N is a countable set. The collection $\{I_j\}_j$ is defined as follows: We may assume that the $I \ne (0,1)$, and consider first those dyadic intervals of length $1/2$ that are contained in I; then—if there is something left of positive measure—we add to the collection all dyadic intervals of length $1/4$ that are contained in the remaining part of I, and so on. If at some stage what is left has no interior, then it is countable, and the interval I has the form $(\frac{m}{2^k}, \frac{n}{2^k})$.

Since any dyadic interval that is contained in I must also be contained in one of the intervals of the family $\{I_j\}_j$, it follows that

$$\mathcal{L}^\downarrow(\triangle(I)) \subset \bigcup_j \mathcal{L}^\downarrow(\triangle(I_j))$$

and therefore

$$|\mathcal{L}^\downarrow(\triangle_I)| \leq \sum_j |\mathcal{L}^\downarrow(\triangle(I_j))| \leq 2 \sum_j |I_j| = 2\,|I|.$$

q.e.d.

5.2 Quasi-Dyadic Decompositions

We now use a theorem of M. Christ [Chr90, Theorem 11], stating that every space of homogeneous type admits a *quasi-dyadic decomposition*. A quasi-dyadic decomposition is given, loosely speaking, whenever, outside from a set of measure zero in W, there is a sequence of nested partitions of W into countably many open sets (called *cubes*), in such a way that

1. if two cubes Q, Q' intersect, then one of them is contained in the other;

2. cubes of generation k are uniformly comparable to balls of radius δ^k (for a certain positive δ less than one);

3. the measure of these cubes is not concentrated too much near their boundary.

The prime example of a quasi-dyadic decomposition is the decomposition of the interval $[0,1]$ into open dyadic intervals. In that case, the null set is the countable set of dyadic rationals, and the number δ is equal to $1/2$.

For a space of homogeneous type of finite diameter, the quasi-dyadic decomposition will assume a particularly simple form. The inclusion relations between these open sets are encoded in a tree; it is convenient to make these dependencies more explicit.

Recall that a tree T with root 0 is endowed with a partial order $\leq$ and that $|x|$ is the generation of a vertex $x \in T$ (relative to the root 0).

Let T be a tree with root 0. Let W be a space of homogeneous type of finite diameter. An **arboreal decomposition** of W over T is a map

$$Q : T \to 2^W$$

such that

5.2. QUASI-DYADIC DECOMPOSITIONS 105

1. $Q(x)$ is an open subset of W for each $x \in T$;

2. $Q(0) = W$;

3. if x is a predecessor of y then $Q(y) \subset Q(x)$;

4. if x and y are different vertices of the same generation then $Q(x)$ and $Q(y)$ are disjoint;

5. $\bigcup_{x:|x|=k} Q(x)$ is a set of full measure in W, for each positive integer k.

A **quasi-dyadic decomposition** of a space of homogeneous type W, $W = (W, \nu, \rho)$, over a tree T (with root 0) is an arboreal decomposition Q of W over T such that for each $x \in T$ there is a ball

$$\mathbb{B}(w_Q(x), r_Q(x))$$

in W with the following properties:

1. There is a δ, $0 < \delta < 1$, such that $r_Q(x) = \delta^{|x|}$ for each $x \in T$;

2. there are constants a_0 and C_1 such that, for all $x \in T$,

$$a_0 \cdot \mathbb{B}(w_Q(x), r_Q(x)) \subset Q(x) \subset C_1 \cdot \mathbb{B}(w_Q(x), r_Q(x));$$

3. there are positive constants η, C_2 such that, for all $x \in T$ and $t > 0$,

$$|\{w \in Q(x) : \rho(w, W \setminus Q(x)) \leq t\delta^{|x|}\}| \leq C_2 t^\eta |Q(x)|,$$

where $\rho(w, E) \stackrel{\text{def}}{=} \inf_{u \in E} \rho(w, u)$ and $|E| \stackrel{\text{def}}{=} \nu(E)$, $E \subset W$.

Of special interest to us is the constant C_1 appearing above, in (2). It is called the **gauge constant** of the quasi-dyadic decomposition Q.

Theorem 5.4 *If $W = (W, \nu, \rho)$ is a space of homogeneous type of finite diameter, then there is a tree T with root and a quasi-dyadic decomposition Q of W over T.*

Proof. The statement is a reformulation of [Chr90, Theorem 11]. We have used the fact that one of the sets Q_α^k (in the notation of [Chr90]) must coincide with W, since the latter has finite diameter, and have rescaled the collection of partitions of W so as to start with level zero ($k = 0$); this rescaling only changes the values of a_0, C_1, C_2. Observe that

Lemma 1.14 implies that the index set I_k, in the notation of [Chr90], must be countable.   q.e.d.

The open sets $Q(x)$ are the **cubes** for the quasi-dyadic decomposition. We say that $Q(x))$ is a cube of generation k if $|x| = k$.

If the space of homogeneous type has infinite diameter, then we need to consider a tree that has a (countably) infinite number of elements for each generation: This means that we choose a root at infinity [Tai87]. The notions of an arboreal decomposition and that of a quasi-dyadic decomposition can be extended without difficulty to the case of a tree with root at infinity: We omit the self-evident reformulation, and state the following

Theorem 5.5 *If $W = (W, \nu, \rho)$ is a space of homogeneous type of infinite diameter, then there is a tree T with root at infinity and a quasi-dyadic decomposition Q of W over T.*

Proof. The statement is again a reformulation of [Chr90, Theorem 11]. q.e.d.

Example 5.6 *The usual decomposition of $\mathbb{R}^n$ into dyadic open cubes is a quasi-dyadic decomposition in the sense of Theorem 5.5.*

We may assume, without loss of generality, that the origin 0 of the tree of a quasi-dyadic decomposition has more than one descendant.

If a vertex $x \in T$ and all its descendants have only one direct descendant, then we say[1] that the vertex x is an **atom**, and that the cube $Q(x)$ is an atom for W.

Remark 5.7 *If $x \in T$ is an atom, then the open set $Q(x)$ reduces to a point. In our applications this possibility does not occur; for NTA domains, which are, in particular, bounded domains in $\mathbb{R}^n$, this fact follows from [HK76, Theorem 5.23, p. 247]). We will therefore assume that our space of homogeneous type does not have atoms.*

Lemma 5.8 *Let Q be a quasi-dyadic decomposition of a space of homogeneous type W over a tree T. Then there is a constant $\epsilon > 0$ such that if $x \in T$ has more than one direct descendant and if y is a direct descendant of x, then*
$$\varepsilon < \frac{|Q(y)|}{|Q(x)|}.$$

---
[1] This terminology is adapted from [Chr90, pp. 612–613].

5.2. QUASI-DYADIC DECOMPOSITIONS

Proof. It follows from the doubling property of the measure ν of W and the fact that the cubes $Q(x)$ of generation k are uniformly comparable to balls of radius δ^k. q.e.d.

Corollary 5.9 *Let Q be a quasi-dyadic decomposition of a space of homogeneous type W over a tree T. Then there is a constant $\epsilon > 0$ such if $x \in T$ has more than one direct descendant, then*

$$0 < \epsilon \leq \frac{|Q(y)|}{|Q(x)|} \leq 1 - \epsilon < 1 \tag{5.4}$$

for each proper descendant y of x. The number of direct descendants of any vertex x is bounded above by an absolute constant.

Let Q be a quasi-dyadic decomposition of W over the tree T. The tree T is called a **coding tree** for W. In fact, the **coding map associated to Q**

$$Q^\bullet : W \longrightarrow bT$$

is a certain map $Q^\bullet$ defined *on a subset of W of full measure*. The image of $w \in W$ by $Q^\bullet$ is denoted

$$Q^\bullet : w \longmapsto w^\bullet,$$

and the coding map $Q^\bullet$ is uniquely determined by the property that

$$Q(x) = \{w \in W : w^\bullet \in E(x)\}$$

for each $x \in T$. In fact, for each $w \in W$ outside a set of measure zero and for each positive integer n, there is only one vertex $w^\bullet(n) \in T$ of generation n, such that

$$w \in Q(w^\bullet(n))$$

and the sequence $\{w^\bullet(n)\}_n$ is a geodesic in T, i.e. a point in bT. We define

$$w^\bullet \stackrel{\text{def}}{=} \{w^\bullet(n)\}_n \in bT$$

for each w in W outside a set of measure zero. In particular, the pushforward of the measure ν by the coding map $Q^\bullet$ is a measure on bT, also denoted $|\cdot|$ and explicitly given by

$$|E(x)| \stackrel{\text{def}}{=} \nu(Q(x))$$

for each $x \in T$.

Let Q be a quasi-dyadic decomposition of a space of homogeneous type W over a tree T. Recall that the notion of stopping-time triangle, given in Section 4.2, was based on the notion of relative weight $||y||_x$ of a descendant y of x with respect to x, given in (4.9). In the present context, the notion of relative weight is defined in terms of the push-forward, via the coding map, of the measure in (W, ν), as follows. Let

$$||y||_x \stackrel{\text{def}}{=} \frac{|Q(y)|}{|Q(x)|} \qquad (5.5)$$

be the the **relative weight** of y with respect to x, for each descendant y of x. The notion of **stopping-time triangle for** W is defined as in (4.10), using the weights given in (5.5); we need only make sure that the origin of the stopping-time triangle has more than one direct descendant. This condition was redundant for the trees considered in Section 3.3, but since our tree T arises from a quasi-dyadic decomposition, it may have vertices that have only one direct descendant. As observed in Remark 5.7, the spaces of homogeneous type that we consider in our applications do not have atoms, and, therefore, each vertex has one descendant that has more than one direct descendant. Therefore, by selecting the first such descendant, we can form a stopping-time triangle without losing any part of the boundary. Now, Corollary (5.9) implies that the bound given in (4.11) also holds in this context, and therefore there are regular twists for any stopping-time triangle. Moreover, given a vertex x having more than one direct descendant, we can select a stopping-time triangle having x as origin and arbitrarily large height.

5.3 The Maximal Decomposition of a Ball

In Theorem 5.12 we will need the following **maximal decomposition of a ball in disjoint cubes**, modulo a null set. This is a generalization of the fact that every open interval in the real line can be decomposed into the disjoint union of dyadic open intervals, modulo a null set (in fact, a countable set). This decomposition is not a decomposition of the Whitney-type, since the intervals in the decomposition can touch the boundary of the ball; however, the argument in the proof closely follows the one used for the proof of the Whitney-type decomposition of an open set in $\mathbb{R}^n$ with respect to the usual dyadic decomposition of $\mathbb{R}^n$ into dyadic cubes; cf. [Ste70, pp. 167–168].

Theorem 5.10 *Let* $W = (W, \nu, \rho)$ *be a space of homogeneous type. Let* $B = \mathbb{B}(w, r)$ *be a ball in* W. *Let* Q *be a quasi-dyadic decomposition of* W *over the coding tree* T *(the root of* T *is at infinity if* $\operatorname{diam}(W)$ *is infinite). Then there is a subset* $T(B) \subset T$ *of* T *such that the collection*

5.3. THE MAXIMAL DECOMPOSITION OF A BALL

$\{Q(x)\}_{x\in T(B)}$ *consists of disjoint cubes whose union is a subset of full measure of* B. *More precisely,*

$$B \setminus N \subset \bigcup_{x \in T(B)} Q(x) \subset B, \tag{5.6}$$

where N is the null set obtained as the union of all null sets appearing in (5) in the definition of arboreal decomposition. In particular,

$$|B| = \sum_{x \in T(B)} |Q(x)|.$$

Proof. The set $\widehat{B} \stackrel{\text{def}}{=} \{x \in T : Q(x) \subset B\}$ is non-empty and it has the property that

$$B \setminus N \subset \cup_{x \in \widehat{B}} Q(x) \subset B.$$

For each $x \in \widehat{B}$, let $\tilde{x}$ be the predecessor of x minimal generation such that $Q(\tilde{x}) \subset B$; observe that $Q(x) \subset Q(\tilde{x}) \subset B$ for each $x \in \widehat{B}$. Let $T(B) \stackrel{\text{def}}{=} \{\tilde{x} : x \in \widehat{B}\}$, and observe that if $Q(x) \neq Q(y)$ for $x, y \in T(B)$, then $Q(x) \cap Q(y) = \emptyset$, since otherwise one would strictly contain the other, say $Q(x) \subset Q(y)$, contradicting the maximality of $Q(x)$. Moreover, $B \setminus N \subset \cup_{x \in T(B)} Q(x) \subset B$. q.e.d.

Remark 5.11 *The quasi-dyadic decomposition is linked to the decomposition of Whitney-type of proper open subsets of W. The former is a sequence of partitions of the whole space W into disjoint cubes; the latter is a decomposition into balls (or cubes) of any proper open subset of W. Actually, the Whitney-type decomposition follows from the quasi-dyadic decomposition. In fact, we saw the version of the Whitney-type decomposition in* [CW71, Théorème 1.3, p. 70] *(valid in any space of homogeneous type), where an open set is decomposed in a collection S of Whitney balls (in the sense of (3.1)), for which*

$$\sum_{B \in S} \chi_B(x) \leq c,$$

where the constant c does not depend on the open set. Recall that in $\mathbb{R}^n$ there is also another version of the Whitney-type decomposition, in which the open set is decomposed, except a set of measure zero, in disjoint open dyadic cubes; see [Ste70, pp.167–168]. *Now, it is easy to see that the latter version also holds in any space of homogeneous type, once we fix a quasi-dyadic decomposition. In this version of the Whitney-type decomposition,* the distance of a cube of generation n from the boundary of the given open set, is comparable to δ^n. The proof is similar to the one given in [Ste70, pp.167–168]. We omit further details, since we will not use this construction.

5.4 Admissible Embeddings

Let D be a space of approach to W. Let $\{\mathcal{G}_\alpha\}_{\alpha \in I}$ be an approach system of dilates of $\mathcal{G}$ for (D, W). Our task is the construction of a new approach family for (D, W) exotic and subordinate with respect to $\{\mathcal{G}_\alpha\}_{\alpha \in I}$. The first tool we shall use is a quasi-dyadic decomposition Q of W over a coding tree T; subordinate to this quasi-dyadic decomposition there will be a special *embedding* of the coding tree in the domain. This approach is suggested by the embedding constructed for the unit disc in Section 5.1. In the applications, the embedding will only be defined on a tent of the tree. An **embedding** of the tent $T(x_0)$ in D is a map

$$q : T(x_0) \to D$$

from the tent $T(x_0)$ in the domain D, where we assume that the vertex x_0 has more than one direct descendant; the tree T that we choose for the embedding is precisely a coding tree for W (associated to a given quasi-dyadic decomposition of W). We need to restrict our attention to a certain class of embeddings, which are well adapted to the given quasi-dyadic decomposition and to the given approach system. Let C_1 be the gauge constant of the given quasi-dyadic decomposition Q of W over T. A map

$$q : T(x_0) \to D \qquad (5.7)$$

of the tent $T(x_0)$ in the domain D is an **admissible embedding** with respect to the quasi-dyadic decomposition Q and the approach system $\{\mathcal{G}_\alpha\}_{\alpha \in I}$ if there exist constants C_1', $C_{1\alpha}$ and $C_{1\alpha}'$ such that

$$C_1 \cdot \mathbb{B}(w_Q(x), r_Q(x)) \subset \mathcal{G}^\downarrow(q(x)) \subset C_1' \cdot \mathbb{B}(w_Q(x), r_Q(x)) \qquad (5.8)$$

and

$$C_{1\alpha} \cdot \mathbb{B}(w_Q(x), r_Q(x)) \subset (\mathcal{G}_\alpha)^\downarrow(q(x)) \subset C_{1\alpha}' \cdot \mathbb{B}(w_Q(x), r_Q(x)) \qquad (5.9)$$

for each $x \in T(x_0)$.

Recall that the shadows $\{\mathcal{G}^\downarrow(z)\}_{z \in D}$ of $\mathcal{G}$ are comparable to a certain family of balls

$$\{\mathbb{B}(w_\mathcal{G}(z), r_\mathcal{G}(z))\}_{z \in D};$$

5.4. ADMISSIBLE EMBEDDINGS

thus the content of the previous definition is that this family of balls is comparable to the one given by the quasi-dyadic decomposition, namely

$$\{\mathbb{B}(w_Q(x), r_Q(x))\}_{x \in T}$$

via the embedding, i.e.

$$\{\mathbb{B}(w_{\mathcal{G}}(q(x)), r_{\mathcal{G}}(q(x)))\}_{x \in T(x_0)} \sim \{\mathbb{B}(w_Q(x), r_Q(x))\}_{x \in T}.$$

An admissible embedding

$$q : T(x_0) \to D$$

will be used to construct an approach family $\mathcal{L}^q$ for (D, W), supported in a subset of full measure of $Q(x_0)$, starting from a particular approach family $\mathcal{L}$ for (T, bT) supported in $E(x_0)$. Observe that given an approach family $\mathcal{L}$ for (T, bT) supported in $E(x_0)$, and an embedding q of the tent $T(x_0)$ in D, there is **the associated approach family $\mathcal{L}^q$ for (D, W)** supported in a subset of full measure in $Q(x_0)$, via the coding map $w \in W \mapsto w^{\cdot} \in bT$: We simply set

$$\mathcal{L}^q(w) \stackrel{\text{def}}{=} \bigcup_{x \in \mathcal{L}(w^{\cdot})} \{z \in D : \mathcal{G}^{\downarrow}(z) \supset \mathcal{G}^{\downarrow}(q(x))\}$$

whenever $w \in Q(x_0)$ belongs to the domain of the coding map. Thus $\mathcal{L}^q(w)$ is the $\mathcal{G}$-completion of $\{q(x)\}_{x \in \mathcal{L}(w^{\cdot})}$. We shall choose the approach family $\mathcal{L}$ and the embedding q in a special way.

Recall that an approach family $\mathcal{L}$ for (T, bT) supported on $E(x_0)$ is defined in terms of a countable collection $\mathcal{T}$ of twisted triangles contained in the tent $T(x_0)$. We need to select the collection of triangles in a certain way. Given two separated triangles Υ_1 and Υ_2, we say that Υ_1 is a **predecessor** of Υ_2, (and Υ_2 is a **descendant** of Υ_1), and we write

$$\Upsilon_1 < \Upsilon_2 \tag{5.10}$$

if the origin of Υ_2 is a descendant of one of the vertices of the basis of Υ_1. A **triangulation** $\mathcal{T}$ of the tent $T(x_0)$ is a collection

$$\mathcal{T} = \{\Upsilon_j\}_j \tag{5.11}$$

of triangles in T that are pairwise separated and such that

1. the vertex x_0 is the origin of one of the triangles in $\mathcal{T}$;
2. the origin 0_Υ of each triangle $\Upsilon \in \mathcal{T}$ belongs to $T(x_0)$;

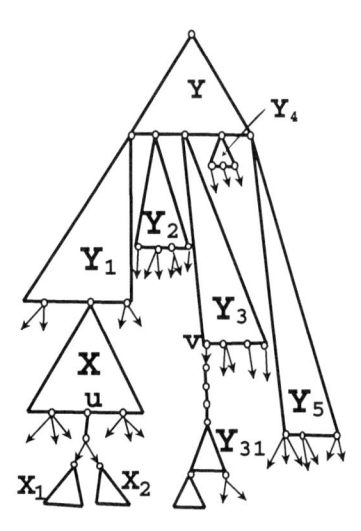

Figure 5.2: The triangles Υ_j are direct descendants of Υ; X_j are direct descendants of X; Υ_{31} is a descendant of Υ_3. Each arrow emanating from a vertex indicates a triangle whose origin is a descendant of the vertex and of minimal generation among those with this property. Thus, only one arrow emanates from the vertex v, but two emanate from the vertex u.

3. for each $\Upsilon \in \mathcal{T}$,

$$E(0_\Upsilon) = \bigcup_{\substack{\Upsilon' \in \mathcal{T} \\ \Upsilon < \Upsilon'}} E(0_{\Upsilon'}).$$

See Fig. 5.2. The meaning of the previous definition is that the collection

$$\{E(0_\Upsilon)\}_{\Upsilon \in \mathcal{T}}$$

is a covering of $E(x_0)$, and therefore $\{Q(0_\Upsilon)\}_{\Upsilon \in \mathcal{T}}$ is a cover of $Q(x_0)$. Moreover, (3) means that *there is no loss in the covering, in passing from one triangle to its descendants*, i.e. that the ball $E(0_\Upsilon) \subset bT$ associated to the origin of any triangle $\Upsilon \in \mathcal{T}$ is the union of the balls associated to descendants of Υ. Observe that the collection $\mathcal{T}$ must be infinite; in particular, every point $\omega \in E(x_0)$ belongs to $E(0_\Upsilon)$ for infinitely many triangles $\Upsilon \in \mathcal{T}$.

We can describe a triangulation $\mathcal{T}$ of a tent $T(x_0)$ by means of a new tree, also denoted by $\mathcal{T}$, whose set of vertices is the collection $\mathcal{T}$ itself. The set $\mathcal{T}$ has a distinguished vertex Υ_0, namely the triangle whose origin coincides with x_0. The vertex $\Upsilon_0 \in \mathcal{T}$ will be used as a root for the tree $\mathcal{T}$. The tree structure in Υ is uniquely determined by the condition that the partial order induced by the distinguished vertex Υ_0 is precisely the

5.4. ADMISSIBLE EMBEDDINGS

one described in (5.10). Observe that, for each triangle $\Upsilon \in \mathcal{T}$, there is a finite set
$$\lambda_+(\Upsilon)$$
of triangles in $\mathcal{T}$, such that the collection $\{E(0_{\Upsilon'}) : \Upsilon' \in \lambda_+(\Upsilon)\}$ is a disjoint cover of $E(0_\Upsilon)$. This follows from (3) in Definition 5.11 and from the compactness of $E(0_\Upsilon)$. The set $\lambda_+(\Upsilon)$ is precisely the set of direct descendants of Υ in the tree $\mathcal{T}$. For each $\Upsilon \in \mathcal{T}$ we denote by $|\Upsilon|$ the **generation** to which Υ belongs **in the tree** $\mathcal{T}$ with root Υ_0, so that $|\Upsilon_0| = 0$, $|\Upsilon| = 1$ for each direct descendant of Υ_0, and so on. We write
$$\Upsilon \to \Upsilon'$$
if Υ' is a direct descendant of Υ in the tree $\mathcal{T}$ with root Υ_0, i.e. if $\Upsilon' \in \lambda_+(\Upsilon)$. In order to avoid too many subscripts, we write
$$E(\Upsilon) \stackrel{\text{def}}{=} E(0_\Upsilon) \subset bT.$$
Recall that each triangle $\Upsilon \in \mathcal{T}$ with basis $\mathcal{B}$ defines the trace maps
$$\Upsilon_\uparrow : E(\Upsilon) \to \mathcal{B}_\Upsilon$$
and
$$\Upsilon_\downarrow : \Upsilon \to 2^{\mathcal{B}_\Upsilon}.$$
The map $\Upsilon_\uparrow$ is uniquely determined by the property
$$\omega \in E(\Upsilon_\uparrow(\omega))$$
for any $\omega \in E(\Upsilon)$; the map $\Upsilon_\downarrow$ is defined as
$$\Upsilon_\downarrow(x) \stackrel{\text{def}}{=} \mathcal{B}_\Upsilon \cap T(x) \subset \mathcal{B}_\Upsilon$$
for each $x \in \Upsilon$. Let $\mathcal{T}$ be a triangulation of the tent $T(x_0)$, and choose a twist σ_Υ of Υ for each triangle $\Upsilon \in \mathcal{T}$. Then **the approach family** $\mathcal{L}$ **for** (T, bT) **supported in** $E(x_0)$ **and spanned by the family**
$$\{(\Upsilon, \sigma_\Upsilon)\}_{\Upsilon \in \mathcal{T}}$$
is defined as
$$\mathcal{L}(\omega) \stackrel{\text{def}}{=} \{x \in T(x_0) : \omega \in E(x)\} \cup \{\sigma_\Upsilon(\Upsilon_\uparrow(\omega)) : \Upsilon \in \mathcal{T}, \omega \in E(\Upsilon)\}$$
for each $\omega \in E(x_0)$. The shadow of $\mathcal{L}$ is then defined with the aid of the map
$$\xi : T(x_0) \to 2^{T(x_0)}$$

defined by
$$\xi(x) \stackrel{\text{def}}{=} \{x\} \cup \{y \in T(x_0) : \exists \Upsilon \in \mathcal{T}, x \in \Upsilon, y \in \mathcal{B}_\Upsilon, \sigma_\Upsilon(y) \in \Upsilon_\downarrow(x)\}$$
via the identity
$$\mathcal{L}^\downarrow(x) = \bigcup_{y \in \xi(x)} E(y).$$
Since
$$w^{\boldsymbol{\cdot}} \in E(y) \iff w \in Q(y)$$
it follows that the shadow $(\mathcal{L}^q)^\downarrow$ of $\mathcal{L}^q$ is given as follows:
$$(\mathcal{L}^q)^\downarrow(z) = \bigcup_{\substack{x \in T \\ \mathcal{G}^\downarrow(q(x)) \subset \mathcal{G}^\downarrow(z)}} \bigcup_{y \in \xi(x)} Q(y).$$

In particular, we see that $\mathcal{L}^q$ is lower semi-continuous.

It is convenient to recast this construction in different terms. We write
$$Q(\Upsilon) \stackrel{\text{def}}{=} Q(0_\Upsilon) \subset W$$
for any triangle $\Upsilon \in \mathcal{T}$, so that the collection of cubes
$$\{Q(\Upsilon) : \Upsilon \in \mathcal{T}, |\Upsilon| = n\}$$
forms a disjoint cover of $Q(\Upsilon_0)$, modulo a null set, for each $n \in \mathbb{N}$. For almost every $w \in Q(x_0)$ (namely, for each $Q(x_0) \setminus N$), there is a unique sequence $\{\Upsilon(w, n)\}_{n \in \mathbb{N}}$ of triangles is $\mathcal{T}$ such that

$$w^{\boldsymbol{\cdot}} \in E(\Upsilon(w, n)), \tag{5.12}$$
$$\Upsilon(w, n) \to \Upsilon(w, n+1) \text{ for } n \in \mathbb{N}, \tag{5.13}$$
$$\Upsilon(w, 0) = \Upsilon_0. \tag{5.14}$$

The origin of $\Upsilon(w, n)$ is denoted by $z(w, n)$. The direct descendant of $z(w, n)$ that belongs to $w^{\boldsymbol{\cdot}}$ is denoted by $z_1(w, n)$. The trace of ω on $\Upsilon(w, n)$ is denoted by $y(w, n)$. The image of $y(w, n)$ by the twist $\sigma_{\Upsilon(w,n)}$ corresponding to $\Upsilon(w, n)$ is denoted by $x(w, n)$. The unique direct descendant of $z(w, n)$ that is a predecessor of $x(w, n)$ is denoted by $z_2(w, n)$. (See Fig. 5.3.) Now, we let

$$s_t(w) \stackrel{\text{def}}{=} \{y(w, n) : n \in \mathbb{N}\} \cup \{x(w, n) : n \in \mathbb{N}\}, \tag{5.15}$$
$$s_g(w) \stackrel{\text{def}}{=} \{q(y(w, n))\}_{n \in \mathbb{N}}, \tag{5.16}$$
$$s_e(w) \stackrel{\text{def}}{=} \{q(x(w, n))\}_{n \in \mathbb{N}}, \tag{5.17}$$
$$s_d(w) \stackrel{\text{def}}{=} s_g(w) \cup s_g(w) \tag{5.18}$$

5.4. ADMISSIBLE EMBEDDINGS

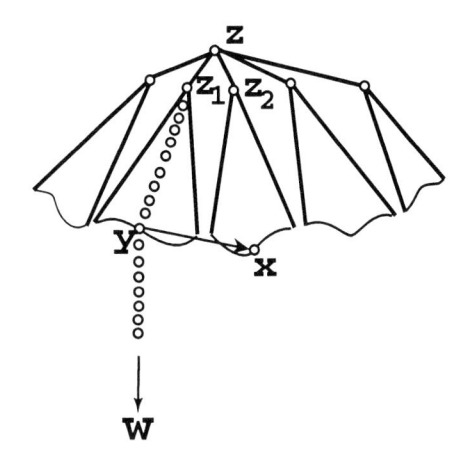

Figure 5.3: The points $x(w,n), y(w,n), z(w,n)$. Also shown are the direct descendants of the origin of the triangle $\Upsilon(w,n)$. The arrow σ denotes the twist $\sigma_{\Upsilon(w,n)}$.

for each $w \in Q_{x_0}$. In particular, for $w \in Q(x_0)$, the set $s_d(w)$ is a sequence in D contained in $\mathcal{L}^q(w)$. Recall that $\tilde{\rho}$ is the approach function $\tilde{\rho}$ for the space of approach D to W. An embedding $q : T(x_0) \to D$ is called **continuous at the boundary** if

$$\lim_{|x| \to \infty} \tilde{\rho}(q(x), w_Q(x)) = 0,$$

where $w_Q(x)$ is the selected center of the ball that is comparable to the cube $Q(x)$ of the given quasi-dyadic decomposition Q. Thus the point $q(x)$ is situated *near* the center of the ball.

Theorem 5.12 *Assume that the following data are given:*

1. *a space of approach $(D, \tilde{\rho})$ to a space of homogeneous type W without atoms;*

2. *a quasi-dyadic decomposition Q of W for the coding tree T;*

3. *an approach system $\{\mathcal{G}_\alpha\}_{\alpha \in I}$ of dilates of $\mathcal{G}$ for (D, W);*

4. *an embedding $q : T(x_0) \to D$ that is admissible and continuous at the boundary $(x_0 \in T)$;*

then there exists an approach family $\mathcal{L}$ for (T, bT) such that the associated $\mathcal{L}^q$ is an approach family for (D, W) supported on a set of full measure in $Q(x_0)$, subordinated to $\mathcal{G}$ and exotic for $\{\mathcal{G}_\alpha\}_\alpha$ on a subset of full measure of $Q(x_0)$.

Proof of Theorem 5.12. We first consider the requirement that the approach family $\mathcal{L}^q$ is subordinated to $\mathcal{G}$. This is equivalent to require that $\mathcal{L}$ satisfies the $\mathcal{G}$-tent condition. We will use Theorem 5.10 to reduce the verification of the tent condition (1) to the case of cubes $Q(x)$. We will need the following

Claim 5.13 *Let $B \subset W$ be a ball in W which is not the whole space W. Let $B = \cup_{x \in T(B)} Q(x)$ be the maximal decomposition of the ball B with respect to the given quasi-dyadic decomposition. Then*

$$(\mathcal{L}^q)^{\downarrow}(\triangle^{\mathcal{G}}(B)) \setminus N \subset \bigcup_{x \in T(B)} S(x) \,, \tag{5.19}$$

where

$$S(x) \stackrel{\text{def}}{=} \{w \in W : Q(x) \supset Q(y) \text{ for some } y \in \mathcal{L}(w^{\cdot})\} \tag{5.20}$$

for each $x \in T$.

Proof of Claim 5.13. If $w \in \mathcal{L}^{q\downarrow}(\triangle^{\mathcal{G}}(B)) \setminus N$ then there is some point $y \in \mathcal{L}(w^{\cdot})$ such that $\mathcal{G}^{\downarrow}(q(y)) \subset B$. Since $\mathcal{G}^{\downarrow}(q(y)) \supset C_1 \cdot Q(y)$, it follows that $Q(y) \subset B$. Now, from the way the maximal decomposition $B = \cup_{x \in T(B)} Q(x)$ of B was constructed, we can select a vertex $x \in T(B)$ such that $Q(y) \subset Q(x)$. **q.e.d.**

Therefore Claim 5.13 reduces the tent condition to the verification of the following

Claim 5.14 *Assume that $\mathcal{L}$ is chosen in such a way that*

$$|S(x)| \lesssim |Q(x)| \tag{5.21}$$

for all $x \in T$. Then the tent condition holds for $\mathcal{L}^q$.

Proof of Claim 5.14. Claim 5.13 and (5.21) imply that if $B = \cup_{x \in T(B)} Q(x)$ is the maximal decomposition of the ball B, then

$$|(\mathcal{L}^q)^{\downarrow}(\triangle^{\mathcal{G}}(B))| \lesssim \sum_{x \in T(B)} |S(x)| \lesssim \sum_{x \in T(B)} |Q(x)| = |B|,$$

which is exactly the tent condition (1). **q.e.d.**

In view of Claim 5.14, we now need to determine conditions on $\mathcal{L}$ such that (5.21) holds. Since $\mathcal{L}$ is determined by the choice of a triangulation $\mathcal{T}$ of $T(x_0)$ by twisted triangles, this means that we only have to describe the sequence $\mathcal{T}$ and the corresponding twists. The triangles will be chosen as stopping-time triangles, defined for the tree T in terms of the relative weights given by (5.5).

5.4. ADMISSIBLE EMBEDDINGS

Claim 5.15 *If we choose the triangulation $\mathcal{T}$ of the tent $T(x_0)$ by twisted triangles in such a way that the triangles in $\mathcal{T}$ are stopping-time triangles relative to W, and the twists are regular, then (5.21) is satisfied.*

Proof of Claim 5.15. Observe that, for a given x, the set $S(x)$ is equal to the union

$$\bigcup_{x \leq y} \bigcup_{z \in \xi(y)} Q(z). \tag{5.22}$$

Now, each descendant y of x is contained in a unique triangle $\Upsilon \in \mathcal{T}$, and either the origin of this triangle is a descendant of x, or it is a predecessor of x. In the first case, the sets $Q(z)$ that appear in (5.22) are contained in $Q(x)$, and therefore their contribution is under control. In the second case, then the triangle Υ contains both x and y: There is only one triangle containing x, and, since the twists are regular,

$$\sum_{z \in \xi(y)} |Q(z)| \lesssim |Q(y)|$$

since the relative weights of (5.5) are now defined with respect to the measure of W. The last term is bounded by $Q(x)$ since $x \leq y$.   q.e.d.

It is a little more work to prove that it is possible to force $\mathcal{L}^q$ to be exotic almost everywhere on $Q(x_0)$. In the following informal discussion we illustrate the main ideas. The plan is to capture the essence of the argument we presented in Section 5.1 in the proof of Lemma 5.2. The argument for Lemma 5.2 was based on the explicit information available for the embedding, and on the idea that the given approach regions behave in an "independent" way at different scales, so that it is extremely improbable to obtain a point at which the approach region is not exotic (we invite the reader to experiment with Figure 5.1). See also [Fef76, p. 57]. In this general setting we do not have the explicit, precise information on the position of the embedded points. A form of the argument based on independence will also appear, based on the condition stated in part (3) of Theorem 5.4, which says that the cubes of the quasi-dyadic decomposition are "round", i.e. their measure is not concentrated near the boundary. A simple estimate, based on Fig. 5.4 and Fig. 5.5, shows that the points at which $\mathcal{L}^q$ is not exotic are close to the boundary of cubes for infinitely many generations. Now, part (3) of Theorem 5.4 implies that, in passing from one generation to the next, the relative weight of points close to the boundary is bounded away from one. This fact forces the set of non-exotic points to have measure zero. Let X_0 be the set of points in Q_{x_0} at which $\mathcal{L}^q$ is not exotic. We need to show that $\nu(X_0) = 0$

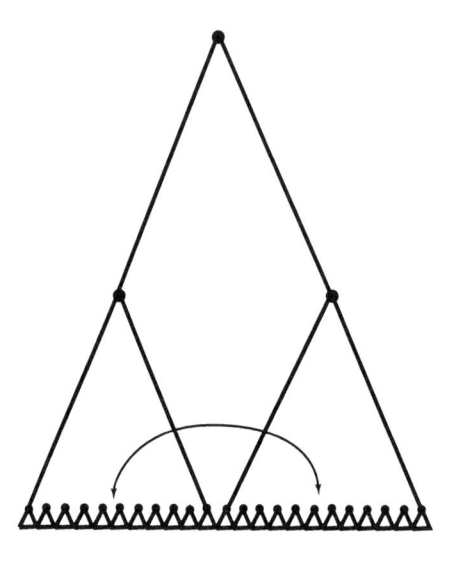

Figure 5.4: The action of a twist

for a certain choice of $\mathcal{L}$. Let $w \in X_0$. Let $\{\zeta_n(w)\}_{n \in \mathbb{N}}$ be the sequence $s_e(w)$, i.e. $\zeta_n(w) = q(x(w,n))$. Then $\{\zeta_n(w)\}_{n \in \mathbb{N}} \subset \mathcal{L}(w)$.

Claim 5.16 *If the embedding q is continuous at the boundary, then*

$$\lim_{n \to \infty} \tilde{\rho}(\zeta_n(w), w) = 0 .$$

Proof of Claim 5.16. Observe that both w and $w_Q(x(w,n))$ belong to $Q(z(w,n))$, whose diameter tends to zero as $n \to \infty$. It follows that $w_Q(x(w,n)) \to w$ as $n \to \infty$, i.e. as $|x(w,n)| \to \infty$. Since q is continuous at the boundary, $\tilde{\rho}(w_Q(x(w,n)), q(x(w,n))) \to 0$ as $n \to \infty$, where $\tilde{\rho}$ is the approach function for (D, W). This fact concludes the proof. **q.e.d.**

Thus, $\lim_n \zeta_n(w) = w$. Therefore, the fact that $\mathcal{L}^q$ is not exotic at w means that there is some $\alpha \in I$ and some $n_1 \in \mathbb{N}$ such that

$$\zeta_n \in \mathcal{G}_\alpha(w) \text{ for each } n \geq n_1,$$

where $\zeta_n \stackrel{\text{def}}{=} \zeta_n(w)$. Let $\{\alpha_k\}_k$ be an increasing sequence of points in I converging to the right-hand point of I. Thus

$$X_0 \subset \bigcup_{k \in \mathbb{N}} \bigcup_{n_1 = 1}^{\infty} C(n_1, \alpha_k),$$

5.4. ADMISSIBLE EMBEDDINGS

where

$$C(n_1, \alpha) = \{w \in Q(x_0) \setminus N : \zeta_n(w) \in G_\alpha(w) \text{ for all } n \geq n_1\} .$$

It is therefore enough to show that $|C(n_1, \alpha_k)| = 0$ for all $n_1 \in \mathbb{N}$ and $k \in \mathbb{N}$.

Claim 5.17 *If $\zeta_n(w) \in G_\alpha(w)$ then*

$$\rho(w, W \setminus Q(z_1(w, n))) < C'_{1\alpha} \delta^{h(\Upsilon(w,n))-1} \delta^{|z_1(w,n)|} ,$$

where $h(\Upsilon)$ is the height of a triangle Υ, defined in (4.7), and $C'_{1\alpha}$ is the constant appearing in (5.9).

Proof of Claim 5.17. Observe that

$$w \in (\mathcal{G}_\alpha)^\downarrow(\zeta_n(w)) = (\mathcal{G}_\alpha)^\downarrow(q(x(w,n))) \subset C'_{1\alpha} \cdot \mathbb{B}(w_Q(x(w,n)), r_Q(x(w,n))),$$

where $\mathbb{B}(w_Q(x(w,n)), r_Q(x(w,n)))$ has radius $\delta^{|x(w,n)|}$. Moreover, $w \in Q(y(w,n)) \subset Q(z(w,n))$. Since $w_Q(x(w,n)) \in Q(x(w,n))$ and the twist $\sigma_{\Upsilon(w,n)}$ is regular, it follows that $w_Q(x(w,n)) \notin Q(z_1(w,n))$, and therefore

$$\rho(w, W \setminus Q(z_1(w,n))) \leq \rho(w, w_Q(x(w,n))) < C'_{1\alpha} r_Q(x(w,n))$$

i.e.

$$\rho(w, W \setminus Q(z(w))) < C'_{1\alpha} \delta^{|x(w,n)|} .$$

Now

$$\delta^{|x(w,n)|} \leq \delta^{h(\Upsilon(w,n))-1} \delta^{|z_1(w,n)|}$$

since

$$|x(w,n)| - |z_1(w,n)| + 1 \geq h(\Upsilon(w,n)) .$$

q.e.d.

Let

$$\tilde{C}_1(n) = \max_{1 \leq j \leq n} C'_{1\alpha_j} , \qquad (5.23)$$

where $C'_{1\alpha}$ is the constant appearing in (5.9). We now use the sequence $\tilde{C}_1(n)$ in order to define a particular triangulation of the tent $T(x_0)$. We assume that, for each triangle Υ in the triangulation $\mathcal{T}$, the following conditions hold:

1. Each triangle is a stopping-time triangle relative to W;
2. $h(\Upsilon_1) < h(\Upsilon_2)$ if Υ_2 is a proper descendant of Υ_1,

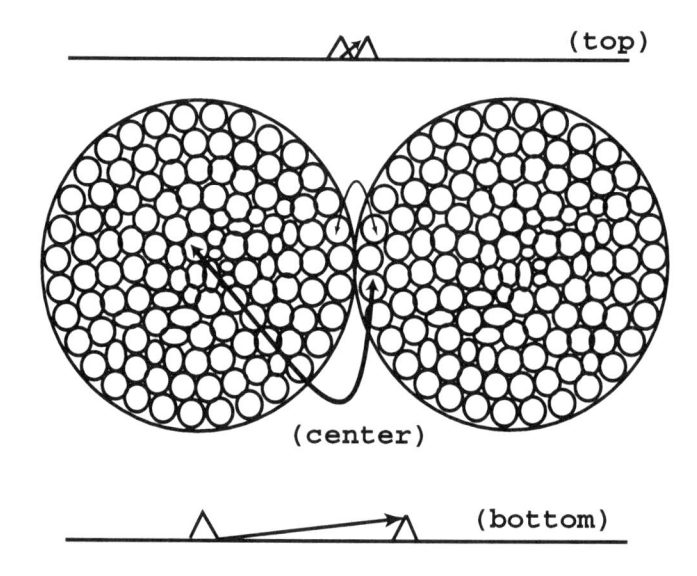

Figure 5.5: In the bottom figure, the "lucky" case, corresponding to the thick arrow in the center picture. In the top figure, the unlucky case, corresponding to the thin arrow in the center picture.

3. $C_2 \cdot [\tilde{C}_1(|\Upsilon|)]^\eta \cdot \delta_1^{h(\Upsilon)-1} \leq 1/2$,

where the $\delta_1 = \delta^\eta$, η, δ and C_2 are the constants given by the quasi-dyadic decomposition, and $|\Upsilon|$ represents the generation to which Υ belongs in the tree $\mathcal{T}$. The possibility of this construction follows from the following facts: (i) We can make the height of a stopping-time triangle as large as we want (therefore, given the generation $|\Upsilon|$ of a triangle, we force the height $h(\Upsilon)$ to be so large that Condition 3 above is satisfied); (ii) the tree T has no atoms, cf. Remark 5.7.

For each triangle Υ in this collection, select a regular twist σ_Υ. Observe that $n = |\Upsilon(w, n)|$.

If $w \in C(n_1, \alpha_k) \setminus N$ and $n \geq n_1$, then $\zeta_n(w) \in \mathcal{G}_{\alpha_k}(w)$, and therefore, by Claim 5.17,

$$\rho(w, W \setminus Q(z_1(w,n))) < C'_{1\alpha_k} \cdot \delta^{h(\Upsilon(w,n))-1} \delta^{|z_1(w,n)|}.$$

Recall that $w \in Q(z_1(w, n))$. These facts lead us to consider the set

$$\Theta(\Upsilon, z^1, \alpha_k) \stackrel{\text{def}}{=} \left\{ w \in Q(z^1) : \rho(w, W \setminus Q(z^1)) < C'_{1\alpha_k} \cdot \delta^{h(\Upsilon)-1} \delta^{|z^1|} \right\},$$

where $\Upsilon \in \mathcal{T}$ and z^1 a direct descendant of the origin of Υ. Claim 5.17 implies that if $w \in C(n_1, \alpha_k)$ and $n \geq n_1$, then

$$w \in \Theta(\Upsilon(w, n), z_1(w, n), \alpha_k).$$

5.4. ADMISSIBLE EMBEDDINGS

Claim 5.18 *If* $\Upsilon \in \mathcal{T}$ *and* $|\Upsilon| \geq k$, *then*

$$|\Theta(\Upsilon, z^1, \alpha_k)| \leq \frac{1}{2}|Q(z^1)|$$

for each direct descendant z^1 of the origin of Υ.

Proof of Claim 5.18. Recall that part (3) in the definition of a quasi-dyadic decomposition implies that

$$|\Theta(\Upsilon, z^1, \alpha_k)| \leq C_2 \cdot t^\eta |Q(z^1)|,$$

where $t \stackrel{\text{def}}{=} C'_{1\alpha_k} \delta^{h(\Upsilon)-1}$. Now, if $|\Upsilon| \geq k$ then $C'_{1\alpha_k} \leq \tilde{C}_1(|\Upsilon|)$, thus

$$C_2 \cdot t^\eta \leq C_2(\tilde{C}_1(|\Upsilon|) \, \delta^{h(\Upsilon)-1})^\eta \leq \frac{1}{2}$$

because of Condition (3) in the choice of the triangulation. **q.e.d.**

Corollary 5.19 *If $n \geq k$ then*

$$|\Theta(\Upsilon(w,n), z_1(w,n), \alpha_k)| \leq \frac{1}{2} \cdot |Q(z_1(w,n))|$$

for each $w \in Q_{x_0}$.

For each triangle $\Upsilon \in \mathcal{T}$, and each α_k, let

$$\Theta(\Upsilon, \alpha_k) \stackrel{\text{def}}{=} \bigcup_{z^1} \Theta(\Upsilon, z^1, \alpha_k),$$

where z^1 runs through the set of direct descendants of the origin 0_Υ of Υ. The set $\Theta(\Upsilon, \alpha_k)$ represents the collection of "bad points" belonging to Q_Υ. In particular, if $w \in C(n_1, \alpha_k)$ and $n \geq n_1$, then $w \in \Theta(\Upsilon(w,n), \alpha_k)$.

The fact that the cubes corresponding to different descendants of 0_Υ form a disjoint cover of $Q(\Upsilon)$ and Claim 5.18 imply that, for every $\Upsilon \in \mathcal{T}$,

$$|\Theta(\Upsilon, \alpha_k)| \leq \frac{1}{2} \cdot |Q(0_\Upsilon)| \quad \text{if} \quad |\Upsilon| \geq k. \tag{5.24}$$

Now let $\Theta^\infty(\Upsilon, \alpha_k)$ be the set

$$\Theta^\infty(\Upsilon, \alpha_k) \stackrel{\text{def}}{=} \{ w \in Q(\Upsilon) : n \geq |\Upsilon| \Rightarrow w \in \Theta(\Upsilon(w,n), \alpha_k) \}$$

of points that are "bad" for each level after Υ. Then (5.24) and the definition of triangulation (Definition 5.11) imply that

$$\nu(\Theta^\infty(\Upsilon, \alpha_k)) = 0 \quad \text{if } \Upsilon \in \mathcal{T} \text{ and } |\Upsilon| \geq k, \tag{5.25}$$

since the measure of $\Theta^\infty(\Upsilon, \alpha_k)$ is at most $(1/2)^j \cdot \nu(Q_\Upsilon)$ for each large j. Moreover, if we let $N \equiv \max\{n_1, k\}$, then $w \in C(n_1, \alpha_k)$ implies that

$$w \in \Theta^\infty(\Upsilon(w, N), \alpha_k)$$

and therefore

$$C(n_1, \alpha_k) \subset \bigcup_{\substack{\Upsilon \in \mathcal{T} \\ |\Upsilon|=N}} \Theta^\infty(\Upsilon, \alpha_k) \ .$$

Observe that (5.25) implies that each set $\Theta^\infty(\Upsilon, \alpha_k)$, where $|\Upsilon| = N$, is a null set, since $|\Upsilon| \geq N \geq k$, and therefore $C(n_1, \alpha_k)$ has measure zero, thus concluding the proof of Theorem 5.12. **q.e.d.**

Chapter 6

Applications

In order to apply Theorem 5.12, we need to show that to a quasi-dyadic decomposition of the boundary of the domain under consideration, we can associate an admissible embedding, continuous at the boundary, with respect to the (natural) approach families defined in Chapter 3. The existence of embeddings with these properties will be shown in this chapter.

In general it is not possible to construct an admissible embedding for the whole tree, but it is sufficient to be able to select a collection $\{T(x)\}_{x \in S}$ of disjoint tents that cover the boundary of the tree (where $S \subset T$), and, corresponding to each tent $T(x)$, an admissible embedding, continuous at the boundary, which produces an approach family $\mathcal{L}_x$ that satisfies the tent condition and is exotic *on the corresponding cube* $Q(x)$, and is *supported in this cube*. The (countable or finite) union $\mathcal{L} = \cup_{x \in S} \mathcal{L}_x$ is exotic almost everywhere on the whole boundary; moreover, functions belonging to the Hardy classes will converge along $\mathcal{L}$ on the boundary, because they converge along $\mathcal{L}_x$ on $Q(x)$ (since each $\mathcal{L}_x$ satisfies the tent condition), $\mathcal{L}_x$ is supported on a set of full measure in $Q(x)$, and $\{Q(x)\}_{x \in S}$ is a disjoint cover of all of the boundary, modulo a null set.

For bounded domains, we simply observe that the union of a finite collection of approach families that satisfy the tent condition has the same property. (If the collection is countable, we would need to assume that the constants appearing in the tent condition form a convergent series).

6.1 Euclidean Half-Spaces

Theorem 6.1 *On the Euclidean half-space $\mathbb{R}^{n+1}_+$ there exists a Nagel–Stein approach family relative to the cones.*

Proof. We saw in Section 1.3 that the natural approach family for $\mathbb{R}^{n+1}_+$ and its system of dilates $\{\Gamma_\alpha\}_{\alpha \in \mathbb{R}}$ are given by

$$\Gamma(x_0) = \{(x,t) \in \mathbb{R}^{n+1}_+ : |x - x_0| < t\} \qquad (6.1)$$

and

$$\Gamma_\alpha(x_0) = \{(x,t) \in \mathbb{R}^{n+1}_+ : |x - x_0| < \alpha t\} \qquad (6.2)$$

for $x_0 \in \mathbb{R}^n$. Fix a quasi-dyadic decomposition Q of $\mathbb{R}^n$, with gauge constants (δ, C_1). Since

$$(\Gamma_\alpha)^\downarrow(x,t) = \alpha \cdot B(x,t)$$

for every $x \in \mathbb{R}^n$ and $t > 0$, if the vertex $x \in T$ corresponds to the ball $B(w_Q(x), r_Q(x))$, we simply let

$$q(x) \stackrel{\text{def}}{=} (w_Q(x), C_1 \cdot r_Q(x)) \in \mathbb{R}^{n+1}_+ .$$

It is clear that q is an admissible embedding on any tent. It is also clear that this embedding is continuous at the boundary, since the factor C_1 that multiplies $r_Q(x)$ is constant. Consider the collection S_0 of vertices in the tree of generation 0. The corresponding tents cover all of the tree, but some vertices in S_0 may have only one direct descendant. For each $x \in S_0$ having only one direct descendant, consider the first descendant $\hat{x}$ of x having more than one direct descendant. The collection $\{\hat{x}\}_{x \in S_0}$ gives the required cover of $\mathbb{R}^n$. We may therefore apply Theorem 5.12.

q.e.d.

6.2 NTA Domains in $\mathbb{R}^n$

Let $D \subset \mathbb{R}^n$ be an NTA domain. Let $W = (bD, \nu, \rho)$ be the boundary of D, where ν is the harmonic measure and ρ the restriction of the Euclidean metric. Let Γ_α be the nontangential approach family defined in Section 3.1.

Theorem 6.2 *If D is an NTA domain in $\mathbb{R}^n$, then there exists a Nagel–Stein approach family with respect to the nontangential approach system $\{\Gamma_\alpha\}_\alpha$.*

Proof. Let Q be a quasi-dyadic decomposition of W with constants (δ, C_1). We show that it is possible to select a finite number of disjoint tents in T, covering the whole boundary, and, for each tent, to construct an admissible embedding of the tent in D, continuous at the boundary.

Let r_0 and $\mathtt{m}$ be the constants involved in the definition of a nontangentially accessible domain D. Select the positive integer n so that the collection S_n of points in T of generation n has the property that the corresponding radius $r_Q(y)$ is less than $\frac{2}{C_1} r_0$ for each $y \in S_n$. Substitute each point $y \in S_n$ having only one direct descendant, with the first descendant having more than one direct descendant, and get a finite collection S of points in T, such that the corresponding cubes $\{Q(x)\}_{x \in S}$ form a disjoint cover of W.

We restrict our attention to the tent $T(y)$ below one of the points $y \in S$. For each descendant x of y, let $\mathbb{B}(w_Q(x), r_Q(x))$ be the corresponding ball in W. Observe that $r_Q(x) < r_Q(y)$.

Recall that for each point $w \in W$ and each $r < r_0$ there is a point $z(w, r)$ in D, whose distance from w and from the boundary W is comparable to r. (See Section 3.1 for the precise definition.)

Let $\alpha_0 = 3\mathtt{m} - 1$, and define $q(x) \stackrel{def}{=} z(w_Q(x), \frac{C_1}{2} r_Q(x))$ for each $x \in T(y)$. A straightforward verification shows that

$$C_1 \cdot \mathbb{B}(w_Q(x), r_Q(x)) \subset (\Gamma_{\alpha_0})^{\downarrow}(q(x)) \subset C_1' \cdot \mathbb{B}(w_Q(x), r_Q(x)), \quad (6.3)$$

where $C_1' = 1 + (3\mathtt{m} - 1) + \frac{1}{2}C_1$. Moreover, for each $\alpha > \alpha_0$,

$$C_1(\alpha) \cdot \mathbb{B}(w_Q(x), r_Q(x)) \subset (\Gamma_\alpha)^{\downarrow}(q(x)) \subset C_1'(\alpha) \cdot \mathbb{B}(w_Q(x), r_Q(x)), \quad (6.4)$$

where $C_1(\alpha) = \frac{1}{2}(\frac{1+\alpha}{\mathtt{m}} - 1)\frac{1}{2}C_1$ and $C_1'(\alpha) = 2(2+\alpha)\frac{1}{2}C_1$. It is also clear that this embedding is continuous to the boundary, since the factor $\frac{C_1}{2}$ is constant and the distance from $z(w_Q(x), \frac{1}{2}C_1 r_Q(x))$ to $w_Q(x)$ is (uniformly) comparable to $\frac{1}{2}C_1 r_Q(x)$. Thus, we can apply Theorem 5.12. **q.e.d.**

6.3 Finite-Type Domains in $\mathbb{C}^2$

Let D be a pseudoconvex domain in $\mathbb{C}^2$, with smooth boundary, of finite type m. Let $W = (\mathrm{b}D, \nu, \rho)$, where ρ is the anisotropic distance introduced in Section 3.2 and ν is the surface measure on $\mathrm{b}D$. Let $\mathcal{A}_\alpha$ be the approach system defined in Section 3.2.

Theorem 6.3 *On each pseudoconvex domain D of finite type in $\mathbb{C}^2$ there is an approach family subordinate and exotic with respect to $\{\mathcal{A}_\alpha\}_\alpha$.*

Let $\mathbb{B}(w,r)$ denote the ball in W with respect to ρ, of center w and radius r. Denote by d the Euclidean distance in $\mathbb{C}^2$, and let $d(z) \stackrel{\text{def}}{=} \inf\{d(w,z) : w \in W\}$. Recall that, for $w \in W$ and a small positive r, we let

$$D(w,r) \stackrel{\text{def}}{=} \inf_{2 \leq k \leq m} \left(\frac{r}{\Lambda_k(w)}\right)^{1/k},$$

where the functions Λ_k are introduced in (3.8) in Section 3.2. We let

$$z(w,r) \equiv (w,r) \equiv w + rN_w,$$

where N_w is the unit *inner* normal vector at w. Moreover, part 3 of Theorem 3.6 implies that there are constants $b_1(\alpha)$ and $b_2(\alpha)$ such that

$$b_1(\alpha) \cdot \mathbb{B}(w, D(w,r)) \subset \mathcal{A}_\alpha^\downarrow(z(w,r)) \subset b_2(\alpha) \cdot \mathbb{B}(w, D(w,r)), \quad (6.5)$$

where $\mathcal{A}_\alpha$ is the natural approach family defined in Section 3.2. Fix a positive ε_0 such that for each point $z \in \mathbb{C}^2$ whose Euclidean distance from W is less than ε_0, there is one and only one point w in W such that $d(w,z) = d(z)$.

Fix a quasi-dyadic decomposition (T,Q) of W, with constants δ, C_1. For each integer n such that $\delta^n < \varepsilon_0$, we can select, for each vertex y in T of generation n, the first descendant $\tilde{y}$ of y having more than one direct descendant. The collection $S = \{\tilde{y} : y \in T, |y| = n\}$ forms a disjoint cover of W, in the sense that the cubes $\{Q_y\}_{y \in S}$ are disjoint and cover all of W. We will now show that it is possible to select n in such a way that, for each $y \in S$, there is an embedding of the tent $T(y)$, that is admissible and continuous at the boundary with respect to $\{\mathcal{A}_\alpha\}_\alpha$. In particular, we need to define a function $q : T(y) \to D_0$, such that

$$C_1 \cdot \mathbb{B}(w_Q(x), r_Q(x)) \subset (\mathcal{A}_1)^\downarrow(q(x)) \quad (6.6)$$

for every $x \in T(y)$. We let $q(x) \stackrel{\text{def}}{=} (w_Q(x), r_1(x))$, so now we need only define $r_1(x)$. Observe that

$$\begin{aligned}(\mathcal{A}_1)^\downarrow(q(x)) &= (\mathcal{A}_1)^\downarrow(w_Q(x), r_1(x)) \supset b_1(1) \cdot \mathbb{B}(w_Q(x), D(w_Q(x), r_1(x))) \\ &= C_1 \cdot \mathbb{B}(w_Q(x), \frac{b_1(1)}{C_1} D(w_Q(x), r_1(x))),\end{aligned}$$

so we only need to select $r_1(x)$ in such a way that

$$r_Q(x) = \frac{b_1(1)}{C_1} D(w_Q(x), r_1(x)),$$

6.3. FINITE-TYPE DOMAINS IN $\mathbb{C}^2$

i.e.
$$\inf_{2\leq k\leq m}\left(\frac{r_1(x)}{\Lambda_k(w_Q(x))}\right)^{1/k} = c\cdot r_Q(x),$$
where $c \equiv \frac{C_1}{b_1(1)}$.

Let
$$\eta_0 \stackrel{\text{def}}{=} \inf_{2\leq k\leq m}\left(\frac{\varepsilon_0}{||\Lambda_k||_\infty}\right)^{1/k},$$
where $||\Lambda_k||_\infty$ is the supremum norm of Λ_k. Then $\eta_0 > 0$ and
$$\eta_0 \leq \inf_{2\leq k\leq m}\left(\frac{\varepsilon_0}{\Lambda_k(w)}\right)^{1/k}$$
for each $w \in W$.

Now select the integer n in such a way that $c\cdot r_Q(x) < \eta_0$, i.e. $\delta^n < \frac{1}{c}\cdot \eta_0$. Then
$$0 < c\cdot r_Q(x) < \eta_0 \leq D(w_Q(x),\varepsilon_0) = \max_{0<r\leq\varepsilon_0} D(w_Q(x),r),$$
and the existence of $r_1(x)$ such that $c\cdot r_Q(x) = D(w_Q(x), r_1(x))$ follows from the fact that $D(w_Q(x), r)$ is continuous and increasing as a function of r, assumes the value 0 at $r = 0$ and the value $D(w_Q(x),\varepsilon_0)$ at the point $r = \varepsilon_0$. (Observe that, since D_0 is of finite type m, at least one of the Λ_k must be different from zero.)

Similar reasoning shows that this embedding is continuous at the boundary, i.e. that $\tilde{\rho}(q(x), w_Q(x)) \to 0$ as $|x| \to \infty$. Since
$$\tilde{\rho}(q(x), w_Q(x)) = r_1(x),$$
it is enough to show that if $|x| \to \infty$ (i.e. $r_Q(x) \equiv \delta^{|x|} \to 0$) then $r_1(x) \to 0$. For suppose not. Then $r_1(x)$ would remain larger than some positive ε (for a subsequence of values of $|x|$ tending to infinity) and therefore $D(w_Q(x), r_1(x))$ would be greater than
$$\inf_{2\leq k\leq m}\left(\frac{\varepsilon}{||\Lambda_k||_\infty}\right)^{1/k},$$
which is a positive number. This contradicts the hypothesis that
$$D(w_Q(x), r_1(x)) \equiv c\cdot r_Q(x) \to 0.$$

Moreover, in view of (6.5) we deduce that
$$\begin{aligned}\mathcal{A}_\alpha^\downarrow(q(x)) &= \mathcal{A}_\alpha^\downarrow(w_Q(x), r_1(x)) \\ &\subset b_2(1)\cdot \mathbb{B}(w_Q(x), D(w_Q(x), r_1(x))) \\ &= b_2(1)\cdot \mathbb{B}(w_Q(x), cr_Q(x)) = c'\mathbb{B}(w_Q(x), r_Q(x)).\end{aligned}$$

The proof that q is an admissible embedding can now be easily concluded, once again using (6.5). **q.e.d.**

6.4 Strongly Pseudoconvex Domains in $\mathbb{C}^n$

Theorem 6.4 *Any strongly pseudoconvex domain in $\mathbb{C}^n$ admits an approach family that is Nagel–Stein relative to the admissible approach system.*

Proof. Let $D \subset \mathbb{C}^n$ be a strongly pseudoconvex domain. Following [NSW81], we may use exactly the same tools and notation adopted in Section 6.3, the only non-formal difference arising from the necessity of considering a collection of $2n - 2$ vector fields $X_1, X_2, \ldots, X_{2n-2}$, where $L_j = \frac{1}{2}(X_j - iX_{J+n})$ form a basis for $T^{(1,0)}(\mathrm{b}D)$.

Observe that the function Λ_2 is bounded away from zero, and therefore the function $D(w, r)$ is of the order of $r^{1/2}$, uniformly in w. Then the embedding q maps the point x into the point $(w_Q(x), (r_Q(x))^2)$ (here we use the same notation of Section 6.3); (6.5) also holds in this case, by Proposition 2, part (i), in [NSW81]. The other details are identical to those given in Section 6.3 and will be omitted. **q.e.d.**

Notes

The possibility of attacking problems in function theory by means of a reduction to a *discrete* problem is one aspect of the interplay between discrete and continuous mathematical structures. The work of R.R. Coifman and R. Rochberg [CR80], R. Rochberg [Roc77], and R. Rochberg and M. Taibleson [RT13] has been a source of inspiration for the present work. See also [GJ82], and [CCdV88], [LS84], [Kai91], [Pom92, p. 148, p. 232, p. 236].

The study of function theory on trees started with the work of J.P. Serre, P. Cartier, P. Sally and M. Taibleson, motivated by questions arising in representation theory and Fourier analysis for p-adic groups and local fields; see [Tai75], [Car73], [Car72], [Car24]. A recent account can be found in [FTN91]. Trees also appear in the paper [Fur70] by H. Furstenberg, where they are used to study certain questions in ergodic theory (the terminology of "section" is borrowed from this paper, but we have used it in a more restrictive sense). In fact, the symbolic dynamics of certain Markov partitions of Jordan curves can be described by a *coding tree*, whose boundary is in correspondence (modulo null sets) with the Jordan curve. Certain questions in the study of the harmonic measure of a domain can then be attacked, since, in this setup, the harmonic measure is equivalent to the *Gibbs measure* on the boundary of the tree. See [Mak90, pp. 44–54] and references therein, in particular the work of L. Carleson [Car85], A. Manning, F. Przytycki [Prz86], [Prz94], [PZ94]. In the words of N.G. Makarov [Mak90, p. 44]:

> The key idea [in the study of properties of the harmonic measure for a class of Jordan domains satisfying certain self-similarity conditions] [...] consists in the construction of a partition of the boundary compatible with the self-similar structure of the domain (until now the partition we used corresponded to the dyadic filtration of the unit circle, which in no way took the geometry of the domain into account).

Another construction of a family of approach regions that is not group invariant is due to P. Sjögren [Sjö85]. It is obtained by exploiting an explicit sequence of nested partitions.

The original proof of Littlewood's Theorem 1.12 is based on a result of Khinchin on Diophantine approximation. An account of the interactions between harmonic analysis, number theory and ergodic theory can be found in [RW95].

The result of M. Christ on the existence of a quasi-dyadic decomposition is independent of the presence of a group acting on the space. In the presence of a group of motions and a group of dilations, R.S. Strichartz proved the existence of a *self-similar tiling*. See [Str94] and references therein, especially [S9].

The domains we have considered so far belong to the category of *rank one spaces*, for which there is essentially only one direction that is nontangential to the boundary. The kind of phenomena that arise in *higher rank* spaces can be seen by considering the domain formed as product of two Euclidean half-planes. The product of the two boundaries is the plane. Then, a point in the plane can be approached by points in the domain along nontangential radii in many essentially different ways, which correspond to all possible eccentricities of the rectangles centered at the point, with sides parallel to the axis. Correspondingly, the basic covering properties that hold for balls in a space of homogeneous type will fail for rectangles of arbitrary eccentricity. See [Ste93], [Sve95], [Sve96a], [Sve96b]. See also [Röna], [Rönb], [Rön97].

For the boundary behaviour of subharmonic functions, see [Zha94], [Zio67] and references therein.

Further developments are given in [CS83], [AB96a], [AB96b], [Aik92], [Aik96], [BS94], [CDS92], [GS], [RS97], [SCS95b], [SCS95a].

List of Figures

1.1	The nontangential approach region for the unit disc	10
1.2	Littlewood's Theorem	15
2.1	The relation between $\mathcal{L}$ and its shadow $\mathcal{L}^{\downarrow}$.	29
2.2	The Carleson tent for the Euclidean half-plane.	36
2.3	The construction of the sequence $\{(x_n, y_n)\}_{n \in \mathbb{N}}$.	47
2.4	The Nagel–Stein approach region is the union of the cones at the points (x_n, y_n).	48
3.1	The unit ball.	62
3.2	The complex tangential directions (left); the normal direction (right).	62
3.3	The admissible approach region for the unit ball. Arrows represent $d(z, u + T_u^c(\mathrm{b}D))$.	63
3.4	The admissible approach region for the unit ball: nontangential in the missing direction, parabolic in the others.	63
3.5	The relative size of the largest embedded analytic disc in the complex tangential direction for the unit ball: $h = r^{1/2}(2-r)^{1/2}$.	63
3.6	The graph on the left is not simply connected, since it has a loop. The vertex X has three adjacent vertices Y, Z and W.	73
3.7	The geodesic from x to z.	74
3.8	The root 0, the predecessor x^- of a vertex x, the descendants of the vertex z.	75
3.9	The "mythical ancestor" ω_0 does not belong to the tree. Generations are parametrized by an integer. Each generation contains infinitely many vertices.	76
3.10	The vertex z is used to compute the Euclidean distance between ω and η.	77
3.11	The tent $T(x)$ below x and the ball $E(x)$ in the boundary.	77

4.1　A section and the corresponding triangle; on the right the schematic picture that suggests the terminology. Observe that by repetition of the given triangle we get one of the three identical pieces that form the boundary of the snowflake. The labeling corresponds to the choices available at each stage (Left, Center, Right). 90

4.2　Another example of a triangle Υ in a tree, also showing the trace of a vertex z on Υ. 90

4.3　A "low resolution picture" of a triangle in a tree. 91

4.4　A schematic picture of the trace of a vertex on a triangle. 91

4.5　The trace of a point in bT on a triangle. 91

4.6　The trace of a vertex on a triangle. 92

4.7　A regular twist. The origin of the triangle has three direct descendants. 95

4.8　Case (b). The black vertices represent $(\xi_n)_*(x)$. 97

5.1　Part of the embedded tree. 100

5.2　The triangles Υ_j are direct descendants of Υ; X_j are direct descendants of X; Υ_{31} is a descendant of Υ_3. Each arrow emanating from a vertex indicates a triangle whose origin is a descendant of the vertex and of minimal generation among those with this property. Thus, only one arrow emanates from the vertex v, but two emanate from the vertex u. 112

5.3　The points $x(w,n), y(w,n), z(w,n)$. Also shown are the direct descendants of the origin of the triangle $\Upsilon(w,n)$. The arrow σ denotes the twist $\sigma_{\Upsilon(w,n)}$. 115

5.4　The action of a twist 118

5.5　In the bottom figure, the "lucky" case, corresponding to the thick arrow in the center picture. In the top figure, the unlucky case, corresponding to the thin arrow in the center picture. 120

Guide to Notation

In $\mathbb{R}^n$ and $\mathbb{C}^n$, the symbol $d(x,y)$ is used to denote the Euclidean distance $|x-y|$ between x and y. If D is a fixed domain in Euclidean space, then $d(z) \stackrel{\text{def}}{=} \inf_{w \in bD} d(z,w)$ denotes the distance from z to the boundary bD of D. In a tree, the symbol $d_e(x,y)$ denotes the Euclidean distance $e^{-|x \wedge y|}$ between x and y, while $|x|$ is the generation of a vertex x in the tree.

In a measure space (W, ν), the symbol $|S|$ is used to denote the measure $\nu(S)$ of a subset S of W. In particular, if $E \subset bT$ is a subset of the boundary of a tree endowed with a very regular random walk, then $|E|$ denotes the hitting distribution $\nu(E)$ of the randow walk. The collection of all subsets of a set S is denoted by 2^S while $\#S$ is the number of elements of a finite set S.

The boundary of a domain D in Euclidean space is denoted by bD. The boundary of a tree T is denoted by bT.

If A and B are two sets, then $A \setminus B$ is the difference set $\{a \in A: a \notin B\}$, while if A and B are subsets of $\mathbb{R}^n$ then $A-B$ is equal to $\{a-b: a \in A, b \in B\}$.

If a and b are positive quantities that depend on certain variables, then the notation
$$a \underset{\sim}{<} b$$
means that there is a *universal* constant c such that $a \leq c \cdot b$ for all values of the variables on which a and b depend; a universal constant c will depend, in general, on certain implicit parameters that are fixed, such as, for example, the particular space under consideration. The notation $a \simeq b$ means than $a \underset{\sim}{<} b$ and $b \underset{\sim}{<} a$.

Bibliography

[AB96a]   H. Aikawa and A.A. Borichev, *Quasiadditivity and measure property of capacity and the tangential boundary behavior of harmonic functions*, Potential theory—ICPT 94 (Kouty, 1994), de Gruyter, Berlin, 1996, pp. 219–227.

[AB96b]   H. Aikawa and A.A. Borichev, *Quasiadditivity and measure property of capacity and the tangential boundary behavior of harmonic functions*, Trans. Amer. Math. Soc. **348** (1996), no. 3, 1013–1030.

[AC92]   M. Andersson and H. Carlsson, *Boundary convergence in non-nontangential and nonadmissible approach regions*, Math. Scand. **70** (1992), 293–301.

[ADBU96]   N. Arcozzi, F. Di Biase, and R. Urbanke, *Approach regions on trees and the unit disc*, J. Reine Angew. Math. **472** (1996), 157–175.

[Aik90]   H. Aikawa, *Harmonic functions having no tangential limits*, Proc. Amer. Math. Soc. **108** (1990), 457–464.

[Aik91]   H. Aikawa, *Harmonic functions and Green potentials having no tangential limits*, J. London Math. Soc. (2) **43** (1991), 125–136.

[Aik92]   H. Aikawa, *Thin sets at the boundary*, Proc. London Math. Soc. (3) **65** (1992), no. 2, 357–382.

[Aik96]   H. Aikawa, *Bessel capacity, Hausdorff content and the tangential boundary behavior of harmonic functions*, Hiroshima Math. J. **26** (1996), no. 2, 363–384.

[Anc87]   A. Ancona, *Negatively curved manifolds, elliptic operators, and the Martin boundary*, Ann. of Math.(2) **125** (1987), 495–536.

[Anc90]   A. Ancona, *Theorie du potentiel sur les graphes et les varietes*, Ecole d'ete de Probabilites de Saint–Flour XVIII—1988, Springer, Berlin, 1990.

[Arn74]   V.I. Arnold, *Les méthodes mathématiques de la mécanique classique*, Éditions MIR, Moscow, 1974.

[AS85]   M.T. Anderson and R. Schoen, *Positive harmonic functions on complete manifolds of negative curvature*, Ann. of Math.(2) **123** (1985), 429–461.

[Bar78]   S.R. Barker, *Two theorems on boundary values of analytic functions*, Proc. Am. Math. Soc. **68** (1978), 54–58.

[BDN91]   A. Boggess, R. Dwilewicz, and A. Nagel, *The hull of holomorphy of a nonisotropic ball in a real hypersurface of finite type*, Trans. Amer. Math. Soc. **323** (1991), 209–232.

[Bog91]   A. Boggess, *CR manifolds and the tangential Cauchy–Riemann complex*, CRC Press, Boca Raton, 1991.

[Bou37]   G. Bourion, *L'ultraconvergence dans les series de Taylor*, Exposés sur la théorie des fonctions, 8, Actualités scientifiques et industrielles, no. 472, Hermann, Paris, 1937.

[BS94]   R. Berman and D. Singman, *Intermittent oscillation and tangential growth of functions with respect to Nagel–Stein regions on a half-space*, Illinois J. Math. **38** (1994), no. 1, 19–46.

[Bur68]   D.L. Burkholder, *Independent sequences with the Stein property*, Ann. Math. Statist. **39** (1968), 1282–1288.

[Bur89]   K. Burdzy, *Geometric properties of 2-dimensional brownian paths*, Probab. Theory Related Fields **81** (1989), no. 4, 485–505.

[Cal50]   A.P. Calderón, *On the behaviour of harmonic functions on the boundary*, Trans. Amer. Math. Soc. **68** (1950), 47–54.

[Car24]   P. Cartier, *Harmonic analysis on trees*, Harmonic analysis on homogeneous spaces, Proc. Sympos. Pure Math. Vol. XXVI, A.M.S., Providence, R.I., 1973, 419–424.

[Car72]   P. Cartier, *Fonctions harmoniques sur un arbre*, Symposia Mathematica Vol. IX (London), Academic Press, 1972, (Convegno di Calcolo delle Probabilità, INDAM, Roma, 1971), 203–270.

[Car73]   P. Cartier, *Géometrie et Analyse sur les arbres*, Séminaire Bourbaki, 24e année, Vol. 1971–72, Lecture Notes in Math. 317, Springer, Berlin, 1973, Exposé n. 407.

[Car82]   L. Carleson, *Estimates of harmonic measures*, Ann. Acad. Sci. Fenn. Ser. A I Math. **7** (1982), 25–32.

[Car85]   L. Carleson, *On the support of harmonic measure for sets of Cantor type*, Ann. Acad. Sci. Fenn. Ser. A I Math. **10** (1985), 113–123.

[Cat89]   D. Catlin, *Estimates of invariant metrics on pseudoconvex domains of dimension two*, Math. Zeit. **200** (1989), 429–466.

[CCdV88] B. Colbois and Y. Colin de Verdier, *Sur la multiplicité de la première valeur propre d'une surface de Riemann à courbure constante*, Comment. Math. Helv. **63** (1988), 194–208.

[CDS92]   P. Cifuentes, J. Dorronsoro, and J. Sueiro, *Boundary tangential convergence on spaces of homogeneous type*, Trans. Amer. Math. Soc. **332** (1992), 331–350.

[Chr90]   M. Christ, *A T(b) theorem with remarks on analytic capacity and the cauchy integral*, Colloq. Math. **60–61** (1990), 601–628.

[CR80]   R.R. Coifman and R. Rochberg, *Representation theorems for holomorphic and harmonic functions in L^p*, Representation theorems for Hardy spaces, Asterisque 77, Soc. Math. France, Paris, 1980, pp. 11–66.

[CS83]   J.L. Cerdà and J. Sueiro, *Approximate identities and convergence at Lebesgue points*, Rend. Circ. Mat. Palermo (2) **32** (1983), no. 1, 5–12.

[CW71]   R.R. Coifman and G. Weiss, *Analyse harmonique non-commutative sur certain espaces homogenes*, Lecture Notes in Math. 242, Springer, Berlin, 1971.

[DB]   F. Di Biase, *Tangential curves and Fatou's theorem on trees*, J. London Math. Soc., to appear.

[dB79] P. de Bartolomeis, *Hardy-like Estimates for the $\bar{\partial}$-Operator and Scripture Theorems for Functions in H^p in Strictly Pseudo-Convex Domains*, Boll. Unione Mat. Ital. (5) **16-B** (1979), 430–450.

[DB95] F. Di Biase, *Approach regions and maximal functions in theorems of Fatou type*, Ph.D. thesis, Washington University, St. Louis, 1995.

[DB97] F. Di Biase, *Exotic convergence in theorems of Fatou type*, Harmonic Functions on Trees and Buildings (Adam Korányi, ed.), Contemporary Mathematics, vol. 206, A.M.S., 1997.

[DBF]   F. Di Biase and B. Fischer, *Boundary behaviour of H^p functions on convex domains of finite type in $\mathbb{C}^n$*, Pacific J. Math., to appear.

[DBFU] F. Di Biase, B. Fischer, and R. Urbanke, *Twist points of the von Koch snowflake*, Proc. Amer. Math. Soc., to appear.

[dG81] M. de Guzmán, *Real variable methods in Fourier analysis*, North-Holland, Amsterdam, 1981.

[DGS96] R. Dror, S. Ganguli, and R.S. Strichartz, *A search for best constants in the Hardy–Littlewood maximal theorem*, J. Fourier Anal. Appl. **2** (1996), no. 5, 473–486.

[DH]   R. Dwilewicz and C.D. Hill, *The normal type function for CR manifolds*, preprint.

[DH92] R. Dwilewicz and C.D. Hill, *An analytic disc approach to the notion of type of points*, Indiana Univ. Math. J. **41** (1992), 713–739.

[Dwi]   R. Dwilewicz, *Pseudoconvexity and analytic discs*, preprint.

[Dyn91] E.M. Dyn'kin, *Methods of the Theory of Singular Integrals: Hilbert Transform and Calderon–Zygmund Theory*, Commutative Harmonic Analysis (V. P. Khavin and N. K. Nikol'skij, eds.), Encyclopaedia Math. Sci. 15, Springer, Berlin, 1991.

[Fat06] P. Fatou, *Séries trigonométriques et séries de Taylor*, Acta Math. **30** (1906), 335–400.

[Fef76]   C.L. Fefferman, *Harmonic analysis and H^p spaces*, Studies in Harmonic Analysis (J.M. Ash, ed.), MAA Stud. Math. 13, Math. Assoc. Amer., Washington, D.C., 1976, pp. 38–75.

[Fol84]   G.B. Folland, *Real analysis–modern techniques and their applications*, Pure and Applied Mathematics, John Wiley & Sons, New York, 1984.

[FS71]   C. Fefferman and E.M. Stein, *Some maximal inequalities*, Amer. J. Math. **93** (1971), 107–115.

[FTN91]   A. Figà-Talamanca and C. Nebbia, *Harmonic analysis and representation theory for groups acting on homogeneous trees*, London Math. Soc. Lecture Note Ser. 162, Cambridge University Press, Cambridge, 1991.

[Fur70]   H. Furstenberg, *Intersections of Cantor sets and transversality of semigroups*, Problems in analysis. A symposium in honor of Salomon Bochner (R.C. Gunning, ed.), Princeton University Press, Princeton, N.J., 1970.

[GCRdF85]   J. Garcia-Cuerva and J.L. Rubio de Francia, *Weighted norm inequalities and related topics*, North-Holland, Amsterdam, 1985.

[GdlH90]   E. Ghys and P. de la Harpe, *Sur les groupes hyperboliques d'aprés Mikhael Gromov*, Birkhäuser, Boston, 1990.

[GJ82]   J.B. Garnett and P.W. Jones, *BMO from dyadic BMO*, Pacific J. Math. **99** (1982), 351–371.

[Gre92]   S. Grellier, *Behaviour of holomorphic functions in complex tangential directions in a domain of finite type in $\mathbb{C}^n$*, Publ. Mat. **36** (1992), 251–292.

[GS]   K. GowriSankaran and D. Singman, *Thin sets and boundary behaviour of the Helmoltz equation*, to appear, Potential Analysis.

[Gun90]   R.C. Gunning, *Introduction to holomorphic functions of several complex variables*, vol. 1, Wadsworth & Brooks/Cole, Belmont, CA, 1990.

[Had92]   J. Hadamard, *Essai sur l'étude des fonctions données par leur développement de Taylor*, J. Math. Pures Appl (4) **8** (1892), 101–186.

[Har06a]   F. Hartogs, *Eine Folgerungen aus der Cauchyschen Integral formel bei Functionen mehrer Veranderlichen*, Munch.Ber. **36** (1906), 223–242.

[Har06b]   F. Hartogs, *Zur Theorie der Analytischen Functionen mehrer unabhangiger Veranderlichen, insbesondere uber die Darstellung derselben durch Reihen welche nach Potenzen einer Veranderlichen fortschreten*, Math.Ann. **62** (1906), 1–88.

[HK76]   W.K. Hayman and P.B. Kennedy, *Subharmonic functions*, Academic Press, London, 1976.

[HL30]   G.H. Hardy and J.E. Littlewood, *A maximal inequality with function-theoretic applications*, Acta Math. **54** (1930), 81–116.

[Hör67]   L. Hörmander, L^p *estimates for (pluri-) subharmonic functions*, Math. Scand **20** (1967), 65–78.

[HS83]   M. Hakim and N. Sibony, *Fonctions holomorphes bornées et limites tangentielles*, Duke Math. J. **50** (1983), 133–141.

[HW]   G.H. Hardy and E.M. Wright, *An introduction to the theory of numbers*, fifth ed., The Clarendon Press, Oxford University Press, New York, 1979.

[JK82]   D. Jerison and C. Kenig, *Boundary behaviour of harmonic functions in non-tangentially accessible domains*, Adv. Math. **46** (1982), 80–147.

[Kai91]   V. Kaimanovich, *Discretization of bounded harmonic functions on Riemannian manifolds and entropy*, Potential Theory (M. Kishi, ed.), Walter de Gruyter & Co., Berlin, 1991.

[Khi24]   A.Ya. Khinchin, *Einige Sätze über kettenbrücke, mit anwendungen auf die Theorie der Diophantischen Approximationen*, Math. Ann. **92** (1924), 115–125.

[Koh72]   J.J. Kohn, *Boundary behaviour of* $\overline{\partial}$ *on weakly pseudoconvex manifolds of dimension two*, J. Diff. Geom. **6** (1972), 523–542.

[Kol25]   A.N. Kolmogorov, *Sur les fonctions harmoniques conjuguées et les séries de Fourier*, Fund. Math. **7** (1925), 24–29.

[Koo93] H. Koo, *Boundary behaviour of holomorphic functions on domains of finite type*, Ph.D. thesis, University of Wisconsin-Madison, 1993.

[Kor65] A. Korányi, *The Poisson integral for generalized half-planes and bounded symmetric domains*, Ann. of Math.(2) **82** (1965), 332–350.

[Kor69] A. Korányi, *Harmonic functions on Hermitian hyperbolic space*, Trans. Amer. Math. Soc. **135** (1969), 507–516.

[KP86] A. Korányi and M.A. Picardello, *Boundary behaviour of eigenfunctions of the Laplace operators on trees*, Ann. Scuola Norm. Sup. Pisa Cl. Sci. (4) **13** (1986), 389–399.

[KPT88] A. Korányi, M.A. Picardello, and M.H. Taibleson, *Hardy spaces on non-homogeneous trees*, Symposia Mathematica Vol. XXIX, Academic Press, New York, 1988, pp. 206–265.

[Kra91] S.G. Krantz, *Invariant metrics and the boundary behaviour of holomorphic functions on domains in $\mathbb{C}^n$*, J. Geom. Anal. **1** (1991), no. 2, 71–97.

[Kra92a] S.G. Krantz, *Function theory of several complex variables*, second ed., Wadsworth & Brooks/Cole, Pacific Grove, CA, 1992.

[Kra92b] S.G. Krantz, *Partial differential equations and complex analysis*, CRC Press, Boca Raton, FL, 1992.

[Leb02] H. Lebesgue, *Intégrale, longuer, aire*, Ann. Mat. Pura Appl.(3) **7** (1902), 231–359.

[Leb04] H. Lebesgue, *Lecons sur l'intégration et la recherche des fonctions primitives*, Gauthier-Villars, Paris, 1904.

[Lem80] L. Lempert, *Boundary behaviour of meromorphic functions of several complex variables*, Acta Math. **144** (1980), 1–26.

[Lev10] E.E. Levi, *Studii sui punti singolari essenziali delle funzioni analitiche di due o piu' variabili complesse*, Ann. Mat. Pura Appl.(3) **17** (1910), 61–87, reprinted in Opere, 1959–60.

[Lev11] E.E. Levi, *Sulle ipersuperficie dello spazio a 4 dimensioni che possono essere frontiera del campo di esistenza di una funzione analitica di due variabili complesse*, Ann. Mat. Pura Appl.(3) **18** (1911), 69–79, reprinted in Opere, 1959–60.

[Lit] J.E. Littlewood, *Collected papers*, The Clarendon Press, Oxford University Press, New York, 1982.

[Lit27] J.E. Littlewood, *Mathematical notes (4): On a theorem of Fatou*, J. London Math. Soc. **2** (1927), 172–176.

[Lit31] J.E. Littlewood, *Mathematical notes (9): on functions subharmonic in a circle (III)*, Proc. London Math. Soc. (2) **32** (1931), 222–234.

[LP] A.J. Lohwater and G. Piranian, *The boundary behaviour of functions analytic in a disk*, Ann. Acad. Sci. Fenn. Ser. A.I. Math. (1957), no. 239.

[LS84] T. Lyons and D. Sullivan, *Function theory, random paths and covering spaces*, J. Diff. Geom. **19** (1984), 299–323.

[Luz13] N.N. Luzin, *Sur la convergence des séries trigonométriques de Fourier*, C. R. Acad. Sci. Paris **156** (1913), 1655–1658.

[Luz16] N.N. Luzin, *Integral i trigonometriceskii ryad. [Integral and Trigonometric Series]*, Mat. Sb. **30** (1916), 1–242 (Russian), dissertation at Moscow University, 1915, [Fortschritte d. Math. **48** (1921/22), p. 1368], [Zbl 45.331], reprinted by G.I.T.T.L., Moscow-Leningrad, 1951 [MR 14, 2g].

[LV73] O. Lehto and K.I. Virtanen, *Quasiconformal mappings in the plane*, second ed., Grundlehren Math. Wiss. 126, Springer, New York, 1973.

[Mak90] N.G. Makarov, *Probability methods in the theory of conformal mappings*, Leningrad Math. J. **1** (1990), 1–56.

[McN94] J.D. McNeal, *Estimates of the Bergman kernel of convex domains*, Adv. Math. **109** (1994), 108–139.

[MPS89a] B.A. Mair, S. Philipp, and D. Singman, *A converse Fatou theorem*, Michigan Math. J. **36** (1989), no. 1, 3–9.

[MPS89b]   B.A. Mair, S. Philipp, and D. Singman, *A converse Fatou theorem on homogeneous spaces*, Illinois J. Math. **33** (1989), no. 4, 643–656.

[MPS90]   B.A. Mair, S. Philipp, and D. Singman, *Generalized local Fatou theorems and area integrals*, Trans. Amer. Math. Soc. **321** (1990), 401–413.

[MS87]   B.A. Mair and D. Singman, *A generalized Fatou theorem*, Trans. Amer. Math. Soc. **300** (1987), no. 2, 705–719.

[Nar85]   R. Narasimhan, *Analysis on real and complex manifolds*, North-Holland, Amsterdam, 1985.

[Nef86]   C.A. Neff, *Maximal function estimates for meromorphic Nevanlinna functions*, Ph.D. thesis, Princeton University, 1986.

[Nef90]   C.A. Neff, *Boundary convergence of functions in the Nevanlinna class*, Colloq. Math. **60** (1990), 477–506.

[NRSW89]   A. Nagel, J.P. Rosay, E.M. Stein, and S. Wainger, *Estimates for the Bergman and Szegö kernels in $\mathbb{C}^2$*, Ann. of Math.(2) **129** (1989), no. 1, 113–149.

[NS84]   A. Nagel and E.M. Stein, *On certain maximal functions and approach regions*, Adv. Math. **54** (1984), 83–106.

[NSW81]   A. Nagel, E.M. Stein, and S. Wainger, *Boundary behavior of functions holomorphic in domains of finite type*, Proc. Nat. Acad. Sci. U.S.A. **78** (1981), no. 11, 6596–6599.

[NSW85]   A. Nagel, E.M. Stein, and S. Wainger, *Balls and metrics defined by vector fields I: Basic properties*, Acta Math. **155** (1985), 103–147.

[Ost26]   A. Ostrowski, *On representation of analytical functions by power series*, J. London Math. Soc **1** (1926), 251–263.

[Poi07]   H. Poincaré, *Les fonctions analytiques de deux variables et la représentation conforme*, Rend. Circ. Mat. Palermo **23** (1907), 185–220.

[Pom92]   Ch. Pommerenke, *Boundary behaviour of conformal maps*, Grundlehren Math. Wiss. 299, Springer, Berlin, 1992.

[Pri]   I.I. Privalov, *Integrale de Cauchy*, Saratov, 1919 (in Russian).

[Pri56]   I.I. Privalov, *Randeigenschaften analytischer funktionen*, second ed., VEB Deutscher Verlag der Wissenschaften, Berlin, 1956.

[Prz86]   F. Przytycki, *Riemann map and holomorphic dynamics*, Invent. Math. **85** (1986), no. 3, 439–455.

[Prz94]   F. Przytycki, *Accessibility of typical points for invariant measures of positive Lyapunov exponents for iterations of holomorphic maps*, Fund. Math. **144** (1994), no. 3, 259–278.

[PZ32]   R.E.A.C. Paley and A. Zygmund, *A note on analytic functions in the unit circle*, Math. Proc. Cambridge. Philos. Soc. **28** (1932), 266–272.

[PZ94]   F. Przytycki and A. Zdunik, *Density of periodic sources in the boundary of a basin of attraction for iteration of holomorphic maps: geometric coding trees technique*, Fund. Math **145** (1994), no. 1, 65–77.

[Rie]   B. Riemann, *Gründlagen für eine allgemeine Theorie der Functionen einer veränderlichen complexen Grösse, (Inauguraldissertation Göttingen 1851)*, Collected papers (R. Narasimhan, ed.), Springer, Berlin, 1990, 35–80.

[Roc77]   R.R. Rochberg, *Decomposition theorems for Bergman spaces and their applications*, Operators and Function Theory (S.C. Power, ed.), Reidel, Dordrecht-Boston, Mass., 1985, 225–277.

[Röna]   J.-O. Rönning, *Convergence for square roots of the Poisson kernel in weakly tangential regions*, Math. Scand., to appear.

[Rönb]   J.-O. Rönning, *A convergence result for square roots of the Poisson kernel in the bidisc*, Math. Scand., to appear.

[Rön97]   J.-O. Rönning, *On convergence for the square root of the Poisson kernel in symmetric spaces of rank* 1, Studia Math. **125** (1997), 219–229.

[RS97]   J.A. Raposo and J. Soria, *Best approach regions for potential spaces*, Proc. Am. Math. Soc. **125** (1997), 1105–1109.

[RT13] R.R. Rochberg and M. Taibleson, *An averaging operator on trees*, Harmonic Analysis and Partial Differential Equations (El Escorial, 1987) (J. Garcia-Cuerva, ed.), Lecture Notes in Math. 1384, Springer, Berlin, 1989, 207–213.

[Rud79] W. Rudin, *Inner function images of radii*, Math. Proc. Cambridge. Philos. Soc. **85** (1979), no. 2, 357–360.

[Rud80] W. Rudin, *Function theory in the unit ball of C^n*, Grundlehren math. Wiss. 241, Springer, New York, 1980.

[Rud88] W. Rudin, *Tangential H^∞-images of boundary curves*, Math. Proc. Cambridge. Philos. Soc. **104** (1988), no. 1, 115–118.

[RW95] J.M. Rosenblatt and M. Wierdl, *Pointwise Ergodic Theorems via Harmonic Analysis*, Ergodic Theory and its Connections with Harmonic Analysis (K.E. Petersen and I.A. Salama, eds.), London Math. Soc. Lecture Note Ser. 205, Cambridge University Press, Cambridge, 1995.

[Sag94] H. Sagan, *Space-filling curves*, Springer, New York, 1994.

[Saw66] S. Sawyer, *Maximal inequalities of weak type*, Ann. of Math. **84** (1966), 157–173.

[Sch72] H.A. Schwarz, *Zur Integration der partiellen Differentialgleichung $\frac{\partial^2 u}{\partial x^2} + \frac{\partial^2 u}{\partial y^2} = 0$*, J. Reine Angew. Math. **74** (1872), 218–253.

[SCS95a] A. Sanchez-Colomer and J. Soria, *A_p and approach regions*, Fourier analysis and partial differential equations., CRC Press, 1995, Proceedings of the conference held in Miraflores de la Sierra, Madrid, Spain, June 15–20, 1992.

[SCS95b] A. Sanchez-Colomer and J. Soria, *Weighted norm inequalities for general maximal operators and approach regions*, Math. Nachr. **172** (1995), 249–260.

[Sjö85] P. Sjögren, *A Fatou theorem and a maximal function not invariant under translation*, Recent Progress in Fourier Analysis (El Escorial, 1983) (I. Peral and J. L. Rubio de Francia, eds.), North Holland, Amsterdam, 1985.

[Sjö86] P. Sjögren, *Admissible convergence of Poisson integrals in symmetric spaces*, Ann. of Math.(2) **124** (1986), 313–335.

[Sjö96]　　P. Sjögren, *Pointwise Convergence for the Square Root of the Poisson Kernel and L^∞ Boundary Functions*, Tech. Report 1996-39, Chalmers University of Technology, 1996, preprint.

[Spi79]　　M. Spivak, *A comprehensive introduction to differential geometry*, vol. 1, Publish or Perish, Wilmington, Del., 1979.

[Ste61]　　E.M. Stein, *On limits of sequences of operators.*, Ann. of Math. **74** (1961), 140–170.

[Ste70]　　E.M. Stein, *Singular integrals and differentiability properties of functions*, Princeton University Press, Princeton, N.J., 1970.

[Ste72]　　E.M. Stein, *Bounday behaviour of holomorphic functions of several complex variables*, Princeton University Press, Princeton, N.J., 1972.

[Ste83]　　E.M. Stein, *Boundary behaviour of harmonic functions on symmetric spaces: maximal estimates for Poisson integrals*, Invent. Math. **74** (1983), 63–83.

[Ste93]　　E.M. Stein, *Harmonic analysis: real-variable methods, orthogonality, and oscillatory integrals*, Princeton University Press, Princeton, N.J., 1993.

[Str94]　　R.S. Strichartz, *Self similarity in harmonic analysis*, J. Fourier Anal. Appl. **1** (1994), no. 1, 1–37.

[Sue86]　　J. Sueiro, *On maximal functions and Poisson–Szegö integrals*, Trans. Amer. Math. Soc. **298** (1986), 653–669.

[Sue87]　　J. Sueiro, *A note on maximal operators of Hardy–Littlewood type*, Math. Proc. Cambridge. Philos. Soc. **102** (1987), 131–134.

[Sue90]　　J. Sueiro, *Tangential boundary limits and exceptional sets for holomorphic functions in Dirichlet-type spaces*, Math. Ann. **286** (1990), 661–678.

[Sue92]　　J. Sueiro, 1992, personal communication.

[SV94]　　M. Salvatori and M. Vignati, *Tangential boundary behaviour of harmonic functions on trees*, Quaderno 16/1994, Università di Milano, 1994.

[Sve95]   O. Svensson, *Extensions of Fatou theorems in products of upper half-spaces*, Math. Scand. (1995), 139–151.

[Sve96a] O. Svensson, *Nonadmissible convergence in symmetric spaces*, J. Reine Angew. Math. (1996), 53–68.

[Sve96b] O. Svensson, *On generalized Fatou theorems for the square root of the Poisson kernel and in rank one symmetric spaces*, Ann. Scuola Norm. Sup. Pisa Cl. Sci. (4) **23** (1996), no. 3, 467–482.

[SW71]   E.M. Stein and G.L. Weiss, *Introduction to Fourier analysis on Euclidean spaces*, Princeton University Press, Princeton, N.J., 1971.

[SZ93]   Y. Sagher and K. Zhou, *On a theorem of Burkholder*, Illinois J. Math. **37** (1993), 637–642.

[Tai75]   M.H. Taibleson, *Fourier Analysis on Local Fields*, Princeton University Press, Princeton, N.J., 1975.

[Tai87]   M.H. Taibleson, *Hardy spaces of Harmonic Functions on Homogeneous Isotropic Trees*, Math. Nachr. **133** (1987), 273–288.

[Thi94]   V. Thilliez, *Caractérisation tangentielle des classes de Carleman de fonctions holomorphes*, Bull. Soc. Math. France **122** (1994), 487–504.

[vK06]   H. von Koch, *Une méthode géométrique élémentaire pour l'étude de certain questions de la théorie des courbes planes*, Acta Math. **30** (1906), 145–174.

[Wei]   K. Weierstrass, *Über das sogenannte Dirichletsche Princip*, Mathematische Werke, Mayer & Müller, Berlin, 1894, read to the Royal Academy of Sciences on July 14, 1870.

[Wie24]   N. Wiener, *Certain notions in potential theory*, J. Math and Phys. **3** (1924), 127–146, reprinted in *Collected works* (P. Masani, ed.), MIT Press, Cambridge, MA, 1976.

[Wit]   R. Wittmann, *A generalized local Fatou theorem*, preprint.

[Zha94]   S. Zhao, *Boundary behavior of subharmonic functions in nontangential accessible domains*, Stud. Math. **108** (1994), no. 1, 25–48.

[Zio67] Ziomek,L., *On the boundary behaviour in the metric L^P of subharmonic functions*, Stud. Math. **29** (1967), 97–105.

[Zyg31] A. Zygmund, *On a theorem of Ostrowski*, J. London Math. Soc. **6** (1931), 162–163.

[Zyg49] A. Zygmund, *On a theorem of Littlewood*, Summa Brasil. Math. **2** (1949), no. 5, 51–57.

[Zyg59] A. Zygmund, *Trigonometric Series*, Cambridge University Press, Cambridge, 1959.

[Zyg76] A. Zygmund, *Notes on the history of Fourier Series*, Studies in Harmonic Analysis (J.M. Ash, ed.), MAA Stud. Math. 13, Math. Ass. Amer., Washington, D.C., 1976, pp. 1–19.

Index

Symbols

$(T, 0)$ 74
0 73
0_Υ 89
2^S 21, 133
$A - B$ 133
$A \setminus B$ 133
$a \lesssim b$ 133
a_0 105
C_1 105
C_2 105
d 7, 57, 58, 61, 126
$d(\cdot)$ 19, 133
$d(\cdot, \cdot)$ 57, 133
d_e 74, 133
d_i 73, 78
E 75, 113
K 59
M' 17
m 125
N 109
N_u 61
$P(z; v)$ 19
Q 114
$Q^{\bullet}$ 107
r_0 57
$r_Q(x)$ 105
$r_{\mathcal{G}}(z)$ 30, 40
$S(\cdot)$ 116
$S_{(x,j)}$ 94
s_d 115
s_e 114
s_g 114
s_t 115
$T(\cdot)$ 75
$T_u^c(\mathrm{b}D)$ 61
$T_u(\mathrm{b}D)$ 60
$w^{\bullet}$ 107
$w_Q(x)$ 105
$w_{\mathcal{G}}(z)$ 30, 40
$x(w, n)$ 114
x^- 74
$y(w, n)$ 114
$z(w, n)$ 114
$z_1(w, n)$ 114
$z_2(w, n)$ 114
$\#$ 94, 133
$\mathcal{A}_\alpha$ 61, 68, 72, 125
$\mathcal{B}_\Upsilon$ 89
$\mathbb{B}(w_Q(x), r_Q(x))$ 105
$\mathbb{B}(w_{\mathcal{G}}(z), r_{\mathcal{G}}(z))$ 30, 40
$\mathbb{B}_W$ 16
$\mathrm{b}D$ 3, 28, 55, 57, 60, 133
$\mathrm{b}T$ 73, 133
$\mathcal{B}$ 89
$z \cdot \overline{w}$ 60
δ 105
diam 16
$\mathcal{L}^*$ 37
$\mathcal{L}^q$ 111
$\mathcal{L}^{\downarrow}$ 28
$\mathcal{L}_\xi$ 92
ε 78
ϵ 106, 107
η 105
ε_0 126
Γ_α 8, 23, 58

$\mathcal{G}^{\downarrow}$ 21
$\mathcal{G}^{\div}$ 37
$\mathcal{G}^{\downarrow}$ 28
Λ 71, 73
$\lambda_+(\Upsilon)$ 113
$\leq$ 74, 75
$\lim_{R \ni z \to w}$ 28
$\lim_{z \to w}$ 28
m........................... 57
$\mathcal{R}^{\downarrow}$ 28
$|\cdot|$ 74, 76, 133
$|\Upsilon|$ 113
ν 17, 58, 60, 62, 72, 78, 105, 107, 124, 125
ν^{z_0} 56
ν_e 32
$\mathcal{P}$ 60
$\pi(\cdot)$ 19, 59, 72
$\mathcal{T}$ 111
$\Theta^{\infty}(\Upsilon, \alpha_k)$ 121
$\Upsilon(\cdot)$ 89
$\Upsilon_1 \leq \Upsilon_2$ 111
$\Upsilon_{(x,\ell)}$ 94
ξ 114
$\|y\|_x$ 93

A
approach families
    union of 28
approach family 28
    associated 111
    complete 39
    completion 39
    exotic 42, 43
    l.s.c. 30
    Nagel–Stein 51
    natural 30
    outer measure 37
    subordinate 32
    translation invariant 43
    twisted triangle 92
approach function 27

approach region
    admissible 61, 68, 72, 125
    cross-section 43
    nontangential ... 8, 23, 58, 78
approach space 27
approach system 40
arboreal decomposition 104
arc 8
atom 106
average operator 37, 80

B
ball 16
    center 17
    radius 17
bounded 17

C
coding map 107
coding tree 107
cone 22, 23, 78, 101
cone condition 43
conjugate harmonic 3
contact manifold 68
convergence
    along approach family 28
    along approach region 28
    in space of approach 28
corkscrew 58
cross-section *see* approach region
cross-section condition 44
cubes 53, 106

D
defining function 66
descendant 74, 111
    direct 74
diameter 16
Dirichlet problem 3
    regular domain 55
distributional inequality 32
domain

defining function 66
finite type 69, 71
Levi pseudoconvex 66
nontangentially accessible 57
NTA 57
of holomorphy 59
smoothly bounded 66
strongly Levi pseudoconvex 67

E
edge 73
embedding 110
    admissible 110
    continuous 115
exterior measure 32

F
family of balls 16
full measure 33

G
gauge constant 105
generation 74, 106
generations
    partition in 75
geodesic 73, 74

H
Hardy space 15, 23
harmonic
    function 3, 18, 73
    measure 56
holomorphic function 3, 59

K
kernel function 59

L
l.s.c.
    approach family 30
labeling 88
Laplace equation 18, 73
Lebesgue point 8

Levi form 66
Littlewood (type) theorem(s) 15, 23, 64

M
maximal
    decomposition .. 11, 103, 108
    function 9, 17, 80
        dyadic 11
    operator 31
meet 74
metric
    anisotropic 62, 69, 71
    Euclidean on trees 74
    hyperbolic 73
missing direction 61, 64, 67

N
Nagel–Stein
    approach family 51

O
operator
    on trees 73
    very regular 76
outer measure 32

P
partial order 74, 75
path 73
Poisson kernel 19
Poisson–Szegö kernel 60
predecessor 74, 111
    direct 74
pseudoconvex
    Levi 66
    strongly Levi 67

Q
quasi-dyadic decomposition . 105
quasi-metric 16

R
random walk 76

regular domain .....*see* Dirichlet problem
regular twist 94
relative weight 108
rescaling 105
root 74
    at infinity 76

S

section 89, 129
    stopping time 94
shadow 21, 28
snowflake .... *see* von Koch
space of homogeneous type ... 17
stopping point 93
supp .... *see* support
support 28

T

tangent space
    complex 61
tangential 79
tangential curve
    on trees 79
tent 36
    on trees 75
    triangulation 111
tent condition 37, 80
trace
    on a triangle 89
tree 72, 73
    coding 129
    dyadic 87
    very regular 76
    with root 74
    with root at infinity 76
triangle 89
    basis 89
    height 92
    interior 89
    origin 89
    stopping time 94

stopping-time—for W ... 108
    twisted 92
triangles
    separated 92
triangulation 111
twist
    regular 94
twist points
    in the von Koch snowflake 58
twisted .... *see* triangle

U

unit disc 3
universal constant 133

V

vertical completion 82
von Koch
    curve 58
    snowflake 58

W

weight
    relative 93
Whitney ball 57

Progress in Mathematics
Edited by:

Hyman Bass
Dept. of Mathematics
Columbia University
New York, NY 10010
USA

J. Oesterlé
Institut Henri Poincaré
11, rue Pierre et Marie Curie
75231 Paris Cedex 05
FRANCE

A. Weinstein
Department of Mathematics
University of California
Berkeley, CA 94720
USA

Progress in Mathematics is a series of books intended for professional mathematicians and scientists, encompassing all areas of pure mathematics. This distinguished series, which began in 1979, includes authored monographs and edited collections of papers on important research developments as well as expositions of particular subject areas.

We encourage preparation of manuscripts in some form of TEX for delivery in camera-ready copy which leads to rapid publication, or in electronic form for interfacing with laser printers or typesetters.

Proposals should be sent directly to the editors or to: Birkhäuser Boston, 675 Massachusetts Avenue, Cambridge, MA 02139, U. S. A.

100 TAYLOR. Pseudodifferential Operators and Nonlinear PDE
101 BARKER/SALLY. Harmonic Analysis on Reductive Groups
102 DAVID. Séminaire de Théorie des Nombres, Paris 1989-90
103 ANGER /PORTENIER. Radon Integrals
104 ADAMS /BARBASCH/VOGAN. The Langlands Classification and Irreducible Characters for Real Reductive Groups
105 TIRAO/WALLACH. New Developments in Lie Theory and Their Applications
106 BUSER. Geometry and Spectra of Compact Riemann Surfaces
107 BRYLINSKI. Loop Spaces, Characteristic Classes and Geometric Quantization
108 DAVID. Séminaire de Théorie des Nombres, Paris 1990-91
109 EYSSETTE/GALLIGO. Computational Algebraic Geometry
110 LUSZTIG. Introduction to Quantum Groups
111 SCHWARZ. Morse Homology
112 DONG/LEPOWSKY. Generalized Vertex Algebras and Relative Vertex Operators
113 MOEGLIN/WALDSPURGER. Décomposition spectrale et séries d'Eisenstein
114 BERENSTEIN/GAY/VIDRAS/YGER. Residue Currents and Bezout Identities
115 BABELON/CARTIER/KOSMANN-SCHWARZBACH. Integrable Systems, The Verdier Memorial Conference: Actes du Colloque International de Luminy
116 DAVID. Séminaire de Théorie des Nombres, Paris 1991-92
117 AUDIN/LaFONTAINE (eds). Holomorphic Curves in Symplectic Geometry
118 VAISMAN. Lectures on the Geometry of Poisson Manifolds
119 JOSEPH/ MEURAT/MIGNON/PRUM/ RENTSCHLER (eds). First European Congress of Mathematics, July, 1992, Vol. I
120 JOSEPH/ MEURAT/MIGNON/PRUM/ RENTSCHLER (eds). First European Congress of Mathematics, July, 1992, Vol. II
121 JOSEPH/ MEURAT/MIGNON/PRUM/ RENTSCHLER (eds). First European Congress of Mathematics, July, 1992, Vol. III (Round Tables)
122 GUILLEMIN. Moment Maps and Combinatorial Invariants of T^n-spaces

123 BRYLINSKI/BRYLINSKI/GUILLEMIN/KAC.
Lie Theory and Geometry: In Honor of
Bertram Kostant
124 AEBISCHER/BORER/KALIN/LEUENBERGER/
REIMANN. Symplectic Geometry
125 LUBOTZKY. Discrete Groups, Expanding
Graphs and Invariant Measures
126 RIESEL. Prime Numbers and Computer
Methods for Factorization
127 HORMANDER. Notions of Convexity
128 SCHMIDT. Dynamical Systems of
Algebraic Origin
129 DIJGRAAF/FABER/VAN DER GEER. The
Moduli Space of Curves
130 DUISTERMAAT. Fourier Integral Operators
131 GINDIKIN/LEPOWSKY/WILSON (eds).
Functional Analysis on the Eve of the
21st Century. In Honor of the Eightieth
Birthday of I. M. Gelfand, Vol. 1
132 GINDIKIN/LEPOWSKY/WILSON (eds.)
Functional Analysis on the Eve of the
21st Century. In Honor of the Eightieth
Birthday of I. M. Gelfand, Vol. 2
133 HOFER/TAUBES/WEINSTEIN/ZEHNDER.
The Floer Memorial Volume
134 CAMPILLO LOPEZ/NARVAEZ MACARRO
(eds) Algebraic Geometry and
Singularities
135 AMREIN/BOUTET DE MONVEL/GEORGESCU.
C_0-Groups, Commutator Methods and
Spectral Theory of N-Body Hamiltonians
136 BROTO/CASACUBERTA/MISLIN (eds).
Algebraic Topology: New Trends in
Localization and Periodicity
137 VIGNERAS. Représentations l-modulaires
d'un groupe réductif p-adique avec $l \neq p$
138 BERNDT/DIAMOND/HILDEBRAND (eds).
139 BERNDT/DIAMOND/HILDEBRAND (eds).
Analytic Number Theory, Vol. 2
In Honor of Heini Halberstam
140 KNAPP. Lie Groups Beyond an
Introduction
141 CABANES. Finite Reductive Groups:
Related Structures and Representations
142 MONK. Cardinal Invariants on
Boolean Algebras
143 GONZALEZ-VEGA/RECIO.
Algorithms in Algebraic Geometry
and Applications

144 BELLAÏCHE/RISLER.
Sub-Riemannian Geometry
145 ALBERT/BROUZET/DUFOUR (eds).
Integrable Systems and Foliations
Feuilletages et Systèmes Intégrables
146 JARDINE. Generalized Etale Cohomology
147 DIBIASE. Fatou TypeTheorems. Maximal
Functions and Approach Regions
148 HUANG. Two-Dimensional Conformal
Geometry and Vertex Operator Algebras
149 SOURIAU. Structure of Dynamical Systems.
A Symplectic View of Physics
150 SHIOTA. Geometry of Subanalytic and
Semialgebraic Sets
151 HUMMEL. Gromov's Compactness Theorem
For Pseudo-holomorphic Curves
152 GROMOV. Metric Structures for Riemannian
and Non-Riemannian Spaces
153 BUESCU. Exotic Attractors: From Liapunov
Stability to Riddled Basins
154 BOTTCHER/KARLOVICH. Carleson Curves,
Muckenhoupt Weights, and Toeplitz
Operators
155 DRAGOMIR. Locally Conformal Kähler
Geometry
156 GUIVARC'H/JI/TAYLOR. Compactifications
of Symmetric Spaces
157 MURTY/MURTY. Non-vanishing of L-
functions and Applications
158 TIRAO/VOGAN/WOLF. Geometry and
Representation Theory of Real and
p-adic Groups